THE WESTRAY TRAGEDY
A Miner's Story

THE WESTRAY TRAGEDY
A Miner's Story

Shaun Comish

Fernwood Publishing
Halifax

Editing: Douglas Beall
Design and layout: Brenda Conroy
Cover photograph: The entrance to the Westray Mine after the May 9, 1992 explosion. Canapress Photo Service.

First printing: August 1993
Second printing: August 1993
Third printing: September 1993
Printed and bound in Canada

A publication of:
Fernwood Publishing
Box 9409, Station A
Halifax, Nova Scotia
B3K 5S3

Canadian Cataloguing in Publication Data

Comish, Shaun.

　　　The Westray Tragedy
　　　ISBN 1 895686 26 1

1. Coal mines and mining — Nova Scotia — Plymouth (Pictou County) 2. Coal mines and mining — Nova Scotia — Plymouth (Pictou County) — Accidents. 3. Westray Mine (N.S.) 4. Westray Mine Disaster, Plymouth, Pictou County, N.S., 1992. I. Title.

TN806.C32N68 1993   363.1'9622334'0971613   C93-098630-X

CONTENTS

ACKNOWLEDGEMENTS

I would like to thank my wife, Shirley, for putting up with all the inconvenience caused to her while I wrote this book. She encouraged me to keep writing when I felt like stopping, and she was a driving force, much more than she may know. I would also like to thank my sister Melanie and my sister-in-law Lynn for their clerical support.

I want to thank Len Bonner for all the help he gave me before and after the explosion. He is a good man, and I wish him all the best in whatever career he pursues.

I would like to thank the family members for the letters and pictures they sent to me; I hope it was a help to them to sit down and write the nice things they had to say about the one they lost.

To Isabel Gillis, I would like to say thank you for all you did; I know it must have been difficult to help out while you had to deal with the loss of your husband, Myles.

Errol Sharpe, the publisher, was a great help in the development of this book and I would like to thank him for his efforts. I would also like to thank Douglas Beall and Brenda Conroy for their superb work in the editing and layout of the book.

INTRODUCTION

My name is Shaun Comish. I worked at the Westray coal mine from September 3rd, 1991, until the May 9th, 1992 explosion, and I was a draegerman during the recovery operation.

I have been mining since 1980, when I started in the small northern Ontario town of Manitouwadge. I spent six years there sharpening my mining skills and knowledge. I attended a Modular Training System in Manitouwadge and a few ground control seminars at the Haileybury School of Mines in Sudbury, Ontario.

I have studied mine rescue for several years and was involved in two mine rescue competitions in Nova Scotia. I started working for Seabright Resources in 1986, where I wrote in-depth training manuals for mining equipment and blasting procedures used in its mines. I took a "Management and Loss Control" course and was a spare supervisor for Seabright. I also worked at several gold mines around Nova Scotia and was a supervisor at the Tangier gold mine. I feel fully qualified to write about the way the Westray mine was run.

The Westray mine and the Foord coal seam lie underground like a large, black spider web. We miners were like insects caught in the trap, occasionally escaping for four days, only to fly back and get caught again. The spider sat and watched, waiting for the right moment. When it finally struck, the wave of horror and grief spread far beyond. This horrific spider now lies in wait once again, ready to strike at those who dare to venture into its black domain.

History has shown us that time does not alter the threat or violence of this spider's attack. The following is a list of previous accidents at the Foord coal seam and mines close by:

1838: explosion, Storage Pit in the Foord Seam—two dead
1858: explosion, Cage Pit in the Foord Seam—two dead
1861: explosion, Bye Pit in the Foord Seam—three dead
1873: explosion, Drummond mine, Westville—sixty dead
1880: explosion, Foord Pit in the Foord Seam—forty-four dead
1885: explosion, Vale mine, Thorburn—thirteen dead
1914: explosion, Allan Shaft in the Foord Seam—two dead
1918: explosion, Allan Shaft—eighty-eight dead
1924: explosion, Allan Shaft—four dead
1935: explosion, Allan Shaft—seven dead
1952: explosion, MacGregor mine, Stellarton—nineteen dead.

The 244 men previously killed should have been reason enough for the company to be extra cautious at Westray. But they were not cautious in any respect as far as I can remember: not about methane, coal dust or even the Coal Mines Regulation Act. As a result, in 1992 the lives of twenty-six more miners were added to the death toll.

As you read this book, it may seem like a tale of mining a hundred years ago, but it is a modern story. This book was not easy to write because it brought back a lot of bad feelings and images I had worked hard to get rid of. However, it probably was good therapy for me to get this off my chest. Many times it looked to me as though there would never be a full inquiry or any charges of consequence laid against the persons responsible and that is what inspired me to write. Someone had to tell what really happened.

My friend Len Bonner started mining in Manitouwadge about the same time I did, and we have worked at four or five different mines together over the years. Lenny is a very safety-conscious worker who watches out not only for himself but everyone he works with. He too is a draegerman who took the "Management and Loss Control" course, took part in the mine rescue competition in Cape Breton and worked at Westray. Len was on the health and safety committee at the mine, so he knows exactly how poorly the company responded to anything reported during safety tours.

There have been many different ideas expressed about what caused the blast of May 9th. People who have never been at the mine site have their versions of the blast and where it started. I,

The author (left) with Lenny Bonner at the Westray mine site. The rock dust to the left was brought in after the explosion.

like a lot of other guys who were in the mine after the explosion, have a very good idea of what happened.

I thought it would be nice if the people who read this book were able to know a bit about each miner who died. I wrote to a member of each of the deceased miners' families, asking if they would like to send a picture and a little something about their loved one. I told them that if it was not what they wanted to do, then I understood and would respect their privacy. I realized it might be difficult for some to write about one so dear to them. The photographs and remembrances that were sent to me are found in the second part of this book, entitled "In Their Memory."

A glossary is provided at the back of the book to help those unfamiliar with coal mining to understand some technical terms I have used while telling this story. Words in the glossary are introduced in italic type in my narrative.

THE WESTRAY TRAGEDY

This is an account of what went on at the Westray mine as told by the men who worked there with me and on the other shifts. I am not a writer by any stretch of the imagination, but I feel the truth must be told.

I started working at Westray on September 3rd, 1991, along with my friends Lenny, Wyman, Stewart, Steve and Rob. When we went in to sign our employment papers, we were told how good the place was and how it had all the most modern equipment to work with. Later we discovered this was true of a lot of the gear but not all. We also found out that all this wonderful equipment doesn't mean much if a mine isn't designed for it.

I am writing this to let you know what it was like working in this hellhole so you can understand the disregard for health and safety that occurred at Westray. This book should also give you insight into what caused the explosion that killed twenty-six of my friends and colleagues. I write this with them always on my mind. Some I knew very well, others not as well, but they were all good men who were stolen from us before their time. I dedicate this work to their memory. I will not forget you, my brothers.

A few years ago, my friend Stewart and I drove up to what is now known as the Westray mine to apply for work with the contractor developing the mine. After we had filled out all the papers, we went upstairs to talk to a guy we had worked for at the Coxheath gold mine. As I was looking out the window at the spot where the portal is today, I thought it would be good to get a job there so I wouldn't have to travel across Canada working at all sorts of mines. When you work for a contractor, this is what usually happens.

Well, we never did get a job with that contractor, so a few years went by before I returned to Pictou County to look for a job. In the meantime, I found work at a mine in Gays River, Nova Scotia, with

The Westray coal mine at Plymouth, Nova Scotia, with New Glasgow and Trenton in the background. The Westray mine was to supply the Nova Scotia Power plant in the distance.

a company now known as Westminer. This was a real pleasant place to work, with a management anybody could get along with. But every good thing eventually comes to an end, and this place was no exception. In May 1991 we got word that the mine was closing down. I sent out my resume to as many places as I could think of but received no reply from any of them.

At the end of August my friend Lenny Bonner called me and asked if I would give him a ride to New Glasgow because he had an interview at Westray. As we drove up, I decided to fill out an application too and, to my surprise, I was hired on the spot. Lenny and I were given an aptitude test we got a good laugh out of, because it was so foolish and simple that if you failed it you probably wouldn't have been able to find your way to the mine site in the first place. I then had to go in and talk with Roger Parry, a man with a strong English accent, while Len talked with the industrial relations man, then vice versa. After we were told when to report for work, we hopped in the car and headed home.

I asked Len if he had understood Roger, because of the way he talked, and Len looked at me and said laughingly, "Not a word he said." On the way home we talked about how nice it was to be getting back to work and how nice this mine sounded. It was a good feeling to think you would be working at the same place for the next fifteen years with no worry of a layoff. Two other guys we had worked with at Westminer had also been hired on and were starting the day we were, so we all would be able to travel together for a while at least. Our prospects of staying in Nova Scotia to work were beginning to look bright.

I had been mining for eleven years, and five of these years in Nova Scotia. Every mine I had worked at in this province had not stayed open long for one reason or another, so I had learned to stay based in one place. Len was a bit more unfortunate than me, because he picked up his belongings and moved to New Glasgow, full of great expectations for a long stay at Westray. I figured I would stay at a rooming house or get an apartment with a couple of other guys from the mine. Travel would then be once every four days, driving eighty miles to get to work. When I had been contracting, I had been away for six weeks at a time, working in a place that you usually had to fly into. The outlook at Westray was pretty good by comparison.

My first day at the mine started at 5:30 am, dragging myself out of bed, then driving all the way to work and back home at night. It was no problem because there were four of us going there from the Dartmouth area, so we were company for each other. That is what we did for the first two weeks.

When we arrived at the mine site on the first day, we were shown where our lockers and hangers were and told to report at deployment after we had changed. Out in deployment we were introduced to Arnie, our fire boss, or supervisor, and then were told that we would be working with John Bates. We went down the main *deep*, or tunnel, about 500 feet, where we were to put up *arches* to support the *roof*. There was a diesel-powered loader called a *scooptram* there to lift the arches, but so many men were working together at this job that we could lift the arches by hand. I couldn't believe that some of the guys would stand between the wall and the "scoop." When a scoop turns, it articulates in the middle, and the front or back will quite often hit the wall. If you were standing there you could be crushed and the operator of the scoop might not even see you.

About noon we stopped for some lunch, and Bates came over and sat with us. The first time I had a conversation with John, I got the feeling he knew how things should be done here, but it was as

Steel arches in a typical coal mine.

A bolter used at Westray.

though he had his hands tied or was told to do things a certain way and never mind about what he thought. John was a lot smarter than he was given credit for. In fact, I think he would have run the mine better and safer. He could get grouchy, but his bark was always worse than his bite. I liked John and will miss the stories he used to tell us at breaktime.

At any rate, on this first day, it seemed everyone wanted to make a good impression because every time anyone in charge asked for something, five guys would run for it. This made for a long day, with too many hands for too few jobs. On the way home we talked about how the main decline was kind of rough, but we figured it was just a temporary thing.

The next day was pretty much the same as the first. About the third day, we were sent down with the mining crews, and we were quite happy about that. Len and I were sent to work on the *bolter* with Roy and Grant. A bolter is a machine used to install roof supports—*rockbolts*, *screening*, or other types of reinforcing. It is an electric-hydraulic, two-boom drilling vehicle that moves on tracks like a bulldozer.

The first incident I experienced at Westray occurred about three days after I began work. Lenny and I were helping on the bolter for the first time, with Roy and Grant. The two of us were on our way back down to the bolter with a couple of screens, used to support the roof, when the two guys on the bolter suddenly disappeared in a cloud of dust. The next thing we could see was a couple of lights shaking at us to stay away. After the dust had cleared, we could see what had happened. The roof had caved in so much that the bolter had to be backed out and the miner brought in to clean up the several tons of rock that had fallen. Lenny and I looked at each other and said, "What have we got ourselves into?"

We had worked together for about eleven years, but this place would prove to be the worst we were ever in. This little display of ground conditions should have told me to leave, but I didn't quit, although I should have. I decided to continue, against my better judgement, and stayed for what would be the worst experience of my life. Days passed and cave-ins became part of everyday life. Sometimes we could work our full four days and not see a single cave-in, but that would only be the rare occasion.

During our first few months, Lenny and I worked on the bolters and saw our share of close calls. Every once in a while a large chunk of roof would come down and destroy something and bring you back to the reality of just how uncertain your life was.

One morning I was told to take a scoop, get some gear and then bring it in closer to where we were working. I was loading the gear into the bucket when I noticed a light at the rear of the scoop, so I went back to see what it was. The mechanic was there fuelling the scoop and I asked him if he wanted me to shut the engine off. He said no, it wasn't necessary, we always leave it running when we work on them. If we shut it off and it doesn't start right away it could be difficult to get compressed air because the pipeline was so far back. I told him it was very dangerous to fuel an engine while it was running and that the fuel was spilling on the ground. I asked where the gear was to clean up the spill, and he just laughed. I figured out then that a lot of fuel was probably mixed with the coal dust throughout the mine.

At every other mine I had worked, the diesel equipment had been fuelled and greased in a designated area and all spills were cleaned up right away. A fire underground is not like a fire on the surface, so you take extra caution to prevent it. The mechanic told me to get used to it because there wasn't any time for anything considered unnecessary. From that point on, I started to see that what he had said was true. Along the sides of the *drift,* or tunnel, I could see half-buried bundles of rockbolts, full cases of grease cartridges, old rags and all sorts of things laying here and there.

One thing I never could get used to was not stopping for lunch at a determined time. Most days on the bolter, the only time you could eat in peace was if one guy left the machine while the other two kept working. Sometimes you had to eat your lunch while working, which meant eating food covered with coal dust. A couple of times we did stop to eat, but as soon as the fire boss or Roger came along, the guys would jump up, close their lunch cans and start working. It didn't matter if we had just stopped for lunch, they jumped up and scurried off anyway. The fact that you worked twelve-hour shifts didn't seem important to management; lunch breaks were frowned on. I guess I must have been considered a rebel, because I told my fire boss that if I didn't eat, I didn't work.

I was starting to get the feeling that this place was quite different from what they had told us at our first interview. Mining has come a long way over the years, but this place was like stepping back in time. The "my way or the highway" attitude was alive and well at Westray.

My first experience on the *shuttle car* was a real "nutbuster." A shuttle car is a thirty-five-foot-long vehicle mounted on rubber tires

Shuttle cars used at Westray, showing (left to right) the front end, the back end with tailgate, and the back end without tailgate.

that is used to haul coal from the *face*, where the mining actually takes place. Chester and Bud were mining up in the Southwest Section and I was moving the trailing cable. Chester asked me if I would like to operate the "car" and, like anybody who wants to learn, I accepted the challenge. Chester showed me all the controls and gave me a brief rundown on how to operate this mammoth machine. I started off in low gear for a while. Then, as I got a little more nerve, I put it into high gear and that is when things started to go awry. The car bounced off the walls and everything else around it. Without any kind of proper training, I had been let free to do unforeseen damage or, worse, injury to somebody who didn't know an untrained person was operating the car.

People were never trained on procedures for walking past an underground vehicle and, as I said, most of the younger, inexperienced guys never thought twice about putting themselves between the wall and a machine. At any rate, I got through the shift without doing too much damage, but I couldn't help wondering why there wasn't some kind of proper training program.

One day I had been working for about nine hours when I got that horrible feeling in my guts which meant I needed to make a trip to the surface for a crap. Going to the surface for this purpose was considered taboo by most of the guys, because the bosses thought you should just go down the drift, away from where you were working and shit along the side of the drift, or go up by the *Stamler* and shovel it onto the conveyor belt to get rid of it. I am a firm believer in being comfortable and clean when I use the toilet, so off I went to surface. If there had been a Johnny-on-the-spot, I would have used it. We had asked for some to be brought down, but that request had been ignored.

As I was leaving, one of the guys on the crew asked me to pick up a few tools we needed. I went out to the main drift and hopped on the first tractor I saw and went up to the surface. At Westray, our tractors were simply farm tractors that had been modified to carry men, light gear and supplies. They were actually illegal underground because of their open exhausts and electric starters. I was about half way back down when I saw the lights of another tractor coming up the ramp, so I pulled into the next *crosscut* to allow him to pass by. As the tractor pulled up, I could see that it was Roger driving, and he didn't look happy. He stopped, looked over at me and said,

"What the fuck is the problem?" I said, "I don't have a problem. I just went to surface to get some tools and to have a shit." At that he started yelling something about me taking his tractor and called me a fucking cunt, but I couldn't make out his jibber-jabber after that because I had put the tractor in gear and taken off down the ramp.

This man was supposed to be a person of professional stature, someone others could look up to for his wisdom and ability to communicate with the men working for him. If all businesses were run like this, there would probably be a lot more of these shootings we hear about in the news.

Not all was bad at Westray. At least I had a pretty decent shift boss. Donnie was the type of guy you could talk to and get your point across to. He would never send you into a situation he wouldn't go into himself, and he was usually pretty easygoing. He liked to see you put in a good day's work but would understand if you slipped up once in a while. Donnie was new at being a boss, but in my opinion he did a rather good job. He could delegate work and discipline workers without causing hard feelings among his crew. Donnie still had a lot to learn, but he is on his way to becoming an excellent supervisor. I wish him all the luck in the world.

One morning in February we were heading up to surface at the end of a shift and it was very cold, as usual. There were seven of us on the tractor, so I figured there was lots of weight to help with traction on the ice that had built up on the ramp. We were progressing up the ramp well when the tractor started to slide sideways and backwards down the ramp. I attempted to go up again, but we just slid again. Most of the guys got off and started walking up, but I knew if I left the tractor stuck, there would be hell to pay, so I tried a few more times to drive up the ramp, to no avail. The only way to get up the ramp was to turn the tractor around and back it up the hill. Although this sounds simple, it was a minus 25° decline with ice all over it and solid rock on either side. It was very risky indeed, but finally I got myself turned around and slowly made my way to surface. Many times people had mentioned that something should be done to heat the decline enough to keep the ice from forming on the roadway and on the roof overhead, but no effective measures were ever taken.

Quite often after cutting a drift, the roof would cave in before you could get in and bolt it. In many cases it kept on caving in until

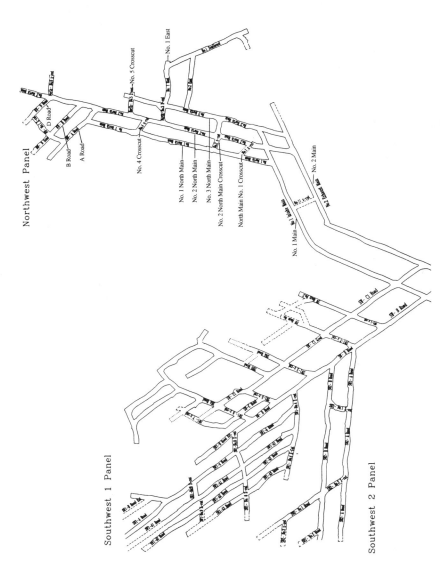

The Southwest Section (left) and the North Mains (right) of the Westray mine.

The rock-cutting head of a typical continuous miner. The continuous miners at Westray had a covered operator's compartment. The roof above is supported with strapping and rockbolts.

it came to a triangular peak, which is known as "churching." This was safer than a half-flat, partly broken roof, because all or most of the stress had been relieved. The height of these areas often reached as high as the original drift.

When bolting an area like this, the only way to get resin into a drilled hole was to stand on top of the canopy, raise the drill platform as high as it would go, and then raise the canopy as high as it would go. So there you are, on top of a canopy designed to protect the man who is drilling from falling rock, and you are standing under open ground with a drop of about twenty-five feet to the rocky floor below. The next trick is to put resin tubes into the hole while trying not to fall. After you do that and get a rockbolt into the hole, you have to climb down off the canopy and down the boom of the drill platform. There were no ladders or handles to hang onto, so you would just climb down like a monkey.

I remember one night we went down and were told to go bolt in

A typical continuous miner transfers coal from front to back on a conveyor belt. The coal is dumped onto a shuttle car (foreground).

No. 2 Main where the back had come down like I have described. We got to the bolter and took a look at the situation, and I decided then and there that I wasn't going to go up on that canopy. Sometimes you just get a feeling that this is not the place to be. The boss came down and looked it over. We had a discussion and I refused to go up on the canopy, so he sent me to work up in No. 1 Main. About two hours later, the guys came up from the bolter and told us the roof had come down and just missed them. They were scared but luckily not hurt.

The pit boss wanted them to go right back down and get back to work. Let me tell you, when a couple of tons of rock just misses you, it makes a guy want to sit down and take it easy for a while before going back to work. One guy told the pit boss that if he was so gung-ho to get it bolted, he could go down and hop to it himself. The pit boss started yelling and screaming like some kind of maniac: "If you don't get back to work, you can go the fuck home!" After this little fit, he walked away mumbling,"Bastard, bastard, bastard." I couldn't help laughing as I thought, *Now there goes a guy who's got it all together.* The bolting crew also laughed at him, probably thinking the same thing I was, then they headed back down to assess the damage.

The cave-in tore out all the screen and most of the rockbolts. Rock was all over the front of the bolter, and large chunks were in the operators' platforms. I am still amazed at how fast you can move when three or four tons of rock are coming at you. The whole time I worked at Westray, only two guys got buried on the bolters, and they were only buried to just above the waist. I know that sounds cold, but there were that many cave-ins that two guys being buried and not hurt too bad is pretty amazing. I swear, if the newspapers or TV had reported every rockfall that occurred, that's all that would have been in the news.

I was involved in quite a few cave-ins. It is difficult to highlight just a few, but one that comes to mind right away is the cave-in at the No. 3 North Main and No. 1 East intersection. John had just brought the miner into the intersection. We noticed it was not working quite right, so the shift boss said he would go and get the mechanic. Just then we noticed a few dribbles of rock falling above the back end of the *continuous miner*. The shift boss said, "That's nothing, it won't go anywhere." I said, "I bet you my pay cheque it's coming down." He laughed and told John to go move the *vent tube* so we could get around the corner with the miner.

He and I stood there for a minute looking at the roof dripping. He then said, "Go give John a hand and I'll go get the mechanic." I went to the front of the miner and he went around the back. Just as I reached the front, I heard the back let go with a sound like ripping wire. The miner started moving up and down like when you push on the back of a car to check the shocks. The roar was deafening as the rock came crashing down on the back end of that miner. I turned and ran downhill. I could see the lights of the guys down the drift working on the bolter, so I signalled them to get the hell out of there. When the first guy got up to where I was, I said, "Come on, I think Donnie is buried under that cave-in because he would have been right under it when it came down." The guy looked at me with disbelief and said, "I don't want to go look." I yelled, "Come on, this is no joke, he was back there when it came down."

We went up beside the miner and looked around back but couldn't see anything because of the dust. We looked behind the miner, but all we could see was a pile of rock, which brought a sickening feeling to my stomach. About a minute later I could see a light coming down the No. 3 North Main. I looked at Donnie in disbelief. He had only been hit with one small rock in the back of the leg as he had run away. This incident could have been very serious indeed if Donnie had not been in good shape and sharp of wit. Luckily the cave-in was not real big, because if it had really let go, the guys down on the bolter would have been trapped in.

Just above this cave-in, about a month later, after we had fixed up the fall and arched it over, we were putting extra arches in the intersection of No. 2 Road and No. 1 East. It was the feeling among most of the men that the only time something was done to support the ground properly was after it had given way. So here we were, trying to support ground that was falling around us as we worked. With a loud crash a piece of rock came down and hit the steel plates just above my head. The next thing I know I'm standing in a cave-in, so I decided real quick to run downhill on No. 2 Road. I only got about thirty feet when I heard it start caving in out in front of me, which left me standing between two rockfalls with nowhere to go. The only thing to do was to stand there and hope the steel sheets above me held up. That time I was lucky, or I wouldn't be telling you this story.

I can remember many times, after being involved in a situation such as I have just described, being quite upset about the whole thing, wondering if I was going to get nailed today, tomorrow or the next day. It was like playing Russian roulette with three bullets in a six-shooter and it was your turn to spin more often than you wanted to.

If you saw the report done by the "Fifth Estate" on the Westray disaster, you will remember Lenny telling Linden MacIntyre about a certain cave-in. He and Wyman were bolting a section that was next in line to be mined, which meant the area had to be ready when the miner was finished cutting the other face. Wyman and Len were working pretty steady to get this face bolted when they heard someone yelling at them. They stopped drilling and Wyman yelled back, asking what was the problem. It was the boss yelling quite excitedly for them to get the fuck out of there. Just after Wyman and Len reached the boss to ask what the worry was, the roof caved in.

Now, again, I'm telling you from experience that when you escape a cave-in, you need to stop for a while and regain your composure. Sometimes it takes a little longer to get up the gall to go back in and start working again. After a couple of minutes of standing there shaking, the boss says, "Come on, guys, get the fuck back to work, we need this place bolted for the miner." Nowadays, when you're presented with a statement like that, most times you shrug your shoulders and go back to work. You know that if you refuse and lose your job, there isn't anything else out there in the way of work.

After several months on the bolter, Len finally got his chance to move on to the mining operations. He started out on the shuttle car, on which he did quite well for a while, then he started on the miner as the second operator. Part of the second operator's job is to make sure the trailing cable is off to the side of the drift so the shuttle car doesn't run over it. The miner doesn't have a cable reel, so you have to move the cable and waterhose out of the way whenever the miner backs up. Len was standing beside the operator's compartment and talking to the operator when the screen above him let go and down came a chunk of the wall. As it broke away, it shattered into pieces before it hit Len. A few pieces hit him in the back and legs and another piece struck him on the back of his head and neck. Only because the miner wasn't running, Len had heard the fall as it started, so he had had a chance to move out the way of the larger chunks that landed where he had been standing.

Len was hurt on his head and neck, so he went right up to the surface without first telling the shift boss. He did this because he knew the boss could sometimes take an hour to find. After he made out an accident report at First Aid, he showered and went to the doctor's. He missed the rest of that shift and the next shift, which was shift four, so he had four days to recover. When he got back to work, he got shit for leaving without notifying the shift boss and was docked in his pay. When Len asked about why he was docked, they told him, "You don't get paid for going home sick." It took a bit of arguing before Len finally convinced the boss that a chunk of coal in the head is not going home sick.

Many times I went back to Len's house—that is where I stayed— and sat there with a coffee, thinking, *I can't do this any more.* After mulling it over for a while, I would go across the street, call home and tell my wife that I was going to quit. She would always be understanding because she knows me well enough to know that I don't scare easy. I have worked in some pretty crazy places, but I had never complained to her before about any place except Westray. She would usually say something like "If it's that bad, quit, come home and we will work it out after." I have never been a quitter, so each time I would think to myself that maybe I was just overtired or cranky and it will probably be better tomorrow. That tomorrow never did come. Now I sit at home and wonder if I could have done something to prevent the terrible loss of my friends' lives.

A couple of times while driving back to Len's house, he and I discussed stopping at the provincial Mines and Energy building and reporting some of the bullshit that was going on at the mine. From what I have heard and seen so far of the Labour Department, I don't think it would have done any good. I truly believe the mine inspectors' hands were tied and their mouths were tightly gagged by some political power.

After each cave-in, we had to clean up the rock that fell, erect arches and shore up the roof with six-by-six timbers, all while working under unbolted roof. Putting up the arches was nerve-wracking because we couldn't hear the ground if it started "working," or loosening and starting to fall again. We used the scooptram to lift the arches and had to stand in the bucket to fasten them together. Many times we found ourselves scrambling out over the back of the bucket to dodge the loose rock falling on us.

Once the arches are up, the fun really begins, because it is time to timber up to the roof with the six-by-six. When *chocking* up, we had to go up on top of the arches to build these timber sets. There we were, about fourteen feet from the ground and standing under unbolted roof. When something started to move on the roof, the options were to jump down into the scoop bucket or to run back into where the arches were already chocked up and hope only a small amount fell. We didn't have much time to think about which option was right, only a second or two.

Another worry was the everpresent methane above the arches. At times the methane level was right off the scale on the handheld meter, which goes up to 5%. This means you are working in an explosive area and the oxygen level has decreased. It is difficult to explain what it's like to be working in an oxygen-deficient atmosphere. You get a ringing in your ears and become lightheaded and kind of confused.

Len told me a story that I found sort of funny at the time, but it could have had a not so funny ending. He and an older guy were above the arches, chocking up down in No. 2 Main. After an hour or so, the chocking was getting pretty high, six or seven feet above the arches. Len was putting chocks in place when he started feeling dizzy and got a buzzing in his ears, so he dropped down on one knee and said, "Man, there is way too much methane up here. Let's get down for a while." The older guy just chuckled and said, "Come on, boy, get out of the way, you'll get used to it." With that the guy grabbed a chock block and climbed up to put it in place. He only got to the top of the timbers when down he came, chock block and all. He landed beside Len on the sheets of metal that cover the arches. When Len looked at him, his eyes were rolled back in his head—he was out like a light. Len brought the guy to and asked if he was all right, but the guy didn't have a clue where he was or what had happened. The thing that made it funny was the way the guy had come down so quick and learned a lesson even quicker. The serious thing about it is that he should have had enough sense to know what would happen. They both got down out of there right away—well, as soon as old buddy got his head together. They then had to run a vent tube up above the arches to suck out the methane. This method works, but as soon as the vent tube is removed, the methane level builds up again and remains built up.

One thing I liked about working above the arches was that the bosses had a hard time getting up to us to bother us with some bullshit way of doing something. I remember one day, three of us were chocking up and Ray, the surveyor, and his partner were working down below. We were getting kind of lightheaded from the methane and were just about to go down for some fresher air when Roger showed up. Needless to say, we decided to stay up above the arches even though we should have gone down. Ray had just set up to shoot lines for the next face to be mined. Roger must had thought he was a surveyor all of a sudden because he started arguing with Ray that it was the wrong way to shoot line. Now, Ray was not to be outdone by Roger, so he started yelling back at him and there began a real good fight. Meanwhile we were up above trying to contain our laughter because it looks like Roger has finally met his match. After a few minutes of steady yelling, Ray grabs the transit, stuffs it into its case and tells Roger, "If you think you're so fucking good, do the job yourself. I quit!" With that he walked away, calling Roger a big pig. It was everything we could do to keep from being heard laughing. This incident put Roger in a real good frame of mind for the rest of the shift, so we ducked out of sight so he would leave without trying to take it out on us.

Roger was one boss you would avoid arguing with because he would never admit he might be wrong. It was futile to try to make him see it your way. It was his way or the highway. For instance, one day Len and three or four other guys were putting a twenty-five-foot steel I-beam up in an area that had a fault running through it. Roger happened along as they were preparing the uprights for the I-beam. After a short conversation, Roger decided the beam had to be brought up the hill and put off to the side of the drift. Len said, "I'll go get the scoop to drag the beam up the hill." When Roger heard him say that, he started yelling at them. He went off the deep end, saying that they didn't need the scoop, they could drag it up the hill by hand. Len couldn't believe what he was hearing, so he said, "You can't be serious!" Of course, Roger lit into him, screaming all sorts of garbage and abuse. In the end they dragged it up the hill by hand. When Len told me this story, I shook my head in disbelief. Why would someone want to risk injury to his men when there was a thirty-five-ton front-end loader right there to do the job? This is just another example of the lack of safety at Westray.

Another example: When we started mining past No. 2 North Main Crosscut, Roger ordered the crews to dive the miner, which meant mining at minus 30° to minus 35°. Our mining equipment is not designed for these steep grades, so naturally we had nothing but trouble the whole time we mined this area. Each time the loaded shuttle car tried to come up the steep grade, we had to chain it to another car and keep pulling until it came up out of the drift.

After the mining was finished, it was a real nightmare to get the miner up out of there. The only way to do it was to come up as far as we could get it until it would start to spin and slide back down, then jack up the miner and put six-by-six timbers under the tracks. Keep in mind that this machine weighs sixty-five tons and is very unpredictable as to which way it will go when you try to move it on these six-by-sixes. I have personally worked for eight to ten hours trying to get the miner up that drift. Every miner operator I talked to couldn't understand why management insisted on trying to mine this drift so steep and why they couldn't design an alternative route. All operators came to the same conclusion: the boss would not admit he had made a mistake. There were a few other areas in the mine where this occurred, but none were as bad as the one I have just described.

One thing I had a hard time with was cave-ins on top of the miner while I was operating it. I can't count how many times cave-ins buried the front of the miner. When all that rock came crashing down, it would shower me with smaller pieces of rock that flew through the protective screening. I would have to duck my head until all the debris stopped flying. Once the dust settled, I would look around and assess the situation. Most times I could back the miner out and clean up the mess, but once in a while I got stuck. When the miner gets stuck, it is usually a real problem—sometimes it could take two shifts to get the miner out. You were always wondering when the roof was going to come crashing down, and it was a real pleasure to mine an area that didn't cave in.

A *dosco* is a mining machine on tracks, with a pineapple-shaped cutting head, that is used to mine through rocky areas. Working with the dosco was a nightmare because we didn't bolt the roof as we advanced. With the dosco we would cut six feet off the face and then put up a set of arches. The big problem was, once again, working under unsupported roof. I hated chocking up to the roof, because several times a large chunk of coal broke off the roof and I had to jump off the cutting head and try to land on my feet on the body of the dosco. After a few dozen times of doing this, our crew got pretty good at this way of roof dodging. Sometimes, when the roof was in better condition, a guy on the ground would throw something against the metal sheets on the sides of the arches just for laughs. This would scare the shit out of the two guys up on the head. It was kind of a crazy thing to do, but it did keep you on your toes.

The worst part of this job was setting the arches in place. Imagine yourself stepping out under unsupported roof to dig down the floor so the arches would sit even. While you're digging, one man is standing back under the arches ready to warn you if a piece starts to fall above you. I remember one time Steve Cyr was digging a footing and Lenny told him to get out. About a second later, a piece six feet long and four feet thick landed right where Steve had been. Again, the only thing you could do when something like this happened was to take five and regain your composure.

We used to complain that we should be bolting as we advanced, but we were told there was no need to bolt when the dosco was mining. Finally, on Christmas Eve we called the boss to the workplace and told him we would not set any more arches without bolting the roof first. He went to see the pit boss and, after we told him the same thing, we were told to go and get the bolter. They were pissed off, but really they couldn't do anything but let us bolt. When I think back on it now, we should have taken a stand a lot sooner, because this was the area of the first major rockfall when the explosion ripped through the mine. When I saw this rockfall later, it didn't surprise me really, because the whole time we worked there you could hear chunks falling on the metal sheets on top of the arches. Sometimes it sounded like a cave-in was starting; I suppose in a sense it was.

It was when we were working in these sorts of conditions that we made pacts with each other. If anything ever happened and one of us died, the other would make sure the reason was found out. Lenny and I talked about this many, many times at work and at home. It was really quite awful being that nervous and worried that you might be killed at work some day.

There were cases of carelessness time after time in that hellhole. Lenny told me of one time when they had brought cutting torches underground to burn holes in some arches. I don't believe one regulation was adhered to when these torches were brought down. Acetylene and oxygen bottles were left lying in the drift, the area where they were burning wasn't watered down, and for rockdusting all they had was guys holding handfuls of the stuff. If the fire got away from where the cutting was going on, these guys would chase the flames and throw rockdust on it. I couldn't believe this was going on, but Len assured me it was. With all the carelessness going on, I'm amazed that what happened in May didn't happen sooner.

One day they were going to burn some holes in the arches where I was working, so I told the boss to come and get me on the surface after they were finished. He asked me if I was scared and I said, "Scared yes, stupid no. If you light those torches, I'm out of here."

As it turned out, for one reason or another, they didn't cut the arches that shift. I can't help wondering how many times we were within inches or seconds of dying. I still can't figure out why I didn't quit a dozen times.

Len came out to the car one day after he had showered up and he was quietly shaking his head as he got in. I asked him what was wrong and what he told me made me realize how close we had come to dying that day. On our four days off, the whole area known now as the old Southwest Section had all started to cave in. This isn't just a small area but quite a significant portion of the mine.

The crews on shift had been busy removing all the gear and equipment before it was all buried. When Len and his crew came on they were told to take the scoop into one of the roads to retrieve a fan. They were getting instructions from the fire boss when Len noticed a mechanic working on the scoop. He looked to see what the mechanic was doing and to his surprise he saw the *methanometer* being disconnected. Seeing this, he said, "What the fuck are you doing?" The answer he got was crazy, just plain crazy: "The scoop will shut off if you don't disconnect it, because the methane levels are too high in the drift where you are going to get the fan." Len tried to explain that the reason the methanometer was on the scoop in the first place is so it can't run in high methane areas, but the reply was, "Don't worry about it."

To this day, Len can't figure out why he hopped on that scoop and drove into that drift knowing what he knew. As the scoop headed into the drift, the motor was making all kinds of weird noises, loud knocking and other sounds Len couldn't describe. The high levels of methane were robbing the engine of the oxygen it needed to run. The methane was obviously above 2.5%, which according to the coal mines Act is too high for men to work in.

Later, as the guys were up in the bucket undoing the chains that held the fan up, the fire boss, who was standing back at the intersection, started shaking his light at them. Underground, when someone shakes their light the way he was, it means get out of where you are as fast as you can. The guys in the bucket jumped down to the ground and ran out ahead of Len. Because Len had to wait to see that everyone was clear of the scoop before he could move it, he found himself in the drift alone. He drove out as fast as he could go and all the time he could see the boss and then a couple of other guys yelling at him to hurry. Just as he got out through the intersection, the whole roof of the intersection caved in right up beyond the existing roof. If the men had not been warned, they would have been trapped in that dead-end drift. By the time gear could have been brought in to dig them out, they would have been dead. The methane was building so quickly that it would have starved them of oxygen in a very short time.

A lot of people ask me why we kept working there. I guess the only answer I can give is that nowadays when you have a job it is very scary to quit and hope to get a job somewhere else. I often felt that maybe things would get better someday. Some guys who worked at Westray didn't really know anything else but mining. That's all they had ever done and probably all they ever will do. The promise of fifteen years of steady work weighed heavy on your mind.

The day I heard about a union trying to organize, I went directly to the hotel where the union guys were and signed a card. I didn't care what union got in, just as long as someone did. With a union, the guys would feel more confident about standing up for themselves. I had seen so many guys get chewed out royally for no good reason sometimes and then not stand up for themselves at all. They knew if they did it could be all over for them. Once the company heard about the union, they gave everybody three new pairs of coveralls and started to tell us about a bonus system that was going to be brought in.

Needless to say, the union vote did not go our way this first time. The failure partly resulted from the way the organizers acted. Buying beers for guys in a bar is a nice gesture if you are spending your own money, but if you're using an expense account, you're spending another local's money. This alone was probably enough to keep some guys from voting for the union to get in, but a little pressure in the right place at the right time by the company sealed its fate. After the union lost the vote, a lot of guys were disappointed, including me. This defeat told the company that things could continue just the way they had been before.

Things did remain just about the same. When the United Steel Workers came to town, Len and I stopped at the Heather Hotel to meet the organizer, Mike Piche, and talk with him. We signed our cards, and Mike asked if we would help him get all the names of the workers and get cards signed. We both agreed to assist in any way we could. We took a bunch of cards and he showed us how to fill them out.

A couple of weeks later I dropped in to see Mike before I headed home for my days off. After we talked for a little, Mike asked me if I would be interested in being on the executive of our local. Because I was heading home, I signed an acceptance to be nominated for any position at all. The meeting was held that night and the next day Mike called to tell me I had been elected interim president. At that point I realized I had a lot of work ahead of me.

I tried to keep my union position quiet at work when I was around staff members. To this day I think I have kept it pretty quiet. I say this because of what happened to a couple of guys who weren't as secretive about their involvement with the union. Although a company isn't allowed to be vindictive towards anyone trying to organize a union, this company did a lot of things they weren't supposed to do.

Lenny relayed what happened to him and I thought, *How typical of these guys.* Len was working underground with two other men up in the Southwest Section. They were chocking up some arches that had been put up on the previous shift. About an hour and a half into the shift, Roger and Gerald Phillips, the general manager of the mine, walked up into the area where Len was working. As soon as Roger came up to them, he started yelling at them, saying they had not done anything yet today. He really started to gross them out, calling them dog fuckers and lazy pricks. Just about then, Lenny said, "Hold it right there! I've had enough of your bullshit. Who the fuck do you think you're talking to, you fat fuck. All you ever do is come around calling people down. You want to get a grip on reality. We have got one whole set of arches chocked up and we've started this next set." Roger barked back, "That was already chocked up by night shift." Len said, "I guess there's no sense talking to you" and started working again. When he turned around to see if Roger had left, he saw him standing there looking at his watch, timing Len to see how long it took him to put up a chock block. Len looked at him and said, "I suppose you're going to stand there like a little kid and time me!" "Yes" was the reply. Len just turned and mumbled, "I'll show you." After a minute, Roger walked away with Gerald and Arnie.

A short time later, Arnie came back and told Len to go to surface and wait to see Gerald. Len asked, "What is all this about?" He was told again to report to surface. Arnie said, "I don't know what it is about, but you better go." So, off went Len, wondering what was happening. When he got to surface, he showered and went out to deployment to have a smoke and wait for Gerald.

Well, an hour went by and then another before Gerald finally showed up. Len had to wait while Gerald showered, and then he was sent up to his office. Gerald talked about what had happened underground and thought that it was maybe a little out of hand. Then he told Len to go and see Roger. When he went into Roger's office and sat down, he started getting ragged on right away. The conversation was more or less one-sided, with Roger going on about how the union wasn't going to save his ass, and he kept saying other stuff about the union. Lenny soon realized what this was all about. Len said, "I never said anything about the union" and that this whole thing was nothing more than intimidation. Roger then said, "You threatened me underground!" Len couldn't believe what he was hearing and said, "You're crazy. I never threatened you at all." With that, Roger growled, "You did and I have two witnesses." Not believing this, Len said, "If you've got witnesses, then bring them in." Roger left and came back in a minute with Gerald. Gerald walked in and said, "That's right, Len, you threatened him. I'm afraid you are suspended until further notice."

Len had to take my car home, so he came back to pick me up after work. When he told me everything that had happened, I suggested we stop and talk to Mike Piche on our way home. We did exactly that, told him the story, and he said, "I'll make a few phone calls and straighten this out." We talked for a while, and then Len and I headed home.

The next day I worked a pretty uneventful day and hurried home to see what had been done to get Len back to work. Len told me that he had gotten a call to go in and see Gerald, and Gerald had told him it was a confusing situation, but he could come back to work with full pay for the time he had missed. Gerald then told him to go see Roger and smooth things over. Len walked into Roger's office, sat down and waited for Roger to say something. After a few minutes, Roger said, "Well, have you decided to smarten up?" Len sat there for a moment and remembered what I had said about letting Roger think he has the upper hand and he will be more reasonable. Len said, "Yeah, I'll smarten up," got up and walked out of the office. The guy that was working with Len when Roger and Gerald came down that day had already been threatened by them, so while all this with Len went on, this other guy never said a word.

I was glad to see Len stand up for himself the way he did. Once the men started to hear what had happened to him and the fact that he got paid for his time off, it showed them that the company did make mistakes and would have to pay when it did. This gave a lot of men more confidence to be themselves and not worry about speaking out once in a while. We, as union organizers, still had a lot of work to do, though. A lot of the workforce were still nervous about signing a card or getting involved. I guess most of them were afraid of being ragged on or given dirty jobs.

Our *flatbed*, or "*boom*," *truck* at Westray was a diesel-operated unit with an open starter and a normal exhaust system. Up until the following incident, it also had normal lights. One day the supply crew was delivering a load of rockbolts and timbers down to the No. 1 Main. While they were unloading these supplies, one of the guys noticed smoke coming from the front end of the flatbed truck. When they ran around to the front end, they saw that the lights had caught fire. Without hesitation, they put it out and took the truck up to the surface. They told the mechanic what had happened and he made his report to the head mechanic. It was only because the lights had caught fire that they were changed to the proper kind of lights. This was a very serious situation that could have had the same results as the events of May 9th. This is just one story of fire underground that didn't get much attention.

Another time was when a guy who had started working at Westray about a month earlier was told to take a scoop to the surface. Now, there are a few differences between the scoops at Westray and other scoops I have operated in the past. All other scoops were air-cooled, but these scoops were water-cooled and their fans blew air out away from the engine. The problem with this is that as you drove up the ramp, the engine was on the up side of the scoop, sending a wicked dust cloud back in your face as the fresh air blew down ramp into the mine. Not only did you get blinded, but coal dust accumulated on the engine. Because of the steep grade of the ramp, the engine would heat up very quickly and get extremely hot.

As this guy was driving to surface, he had to keep clearing his eyes of dust. Clearing his eyes this one time, he noticed flames coming out of the engine compartment. First you think you are seeing things, then you get scared. Once he realized what it was, he set off the fire suppression system, which put out the fire. When he arrived on surface, he reported what had happened and was understandably upset. He did a lot of yelling and complaining, but it was treated like nothing had really happened. This guy must have thought the management were all retarded not to concern themselves with the fact that a piece of equipment had caught fire underground.

Another time a fire occurred was when a crew was driving a tractor up the main decline at the end of a shift. The tractor would heat up very quickly because of its heavy load and the steep grade it had to climb. Quite often the exhaust pipe would get red hot, like the burner on a stove, and sparks would fly out of it. On this particular trip up, there were sparks, and a couple of times the men on the tractor saw some of the sparks ignite bits of coal dust. If you were at a rock concert this would have looked pretty neat, but underground in a coal mine it looks pretty frightening.

These stories of fires underground were not officially reported to the provincial Department of Mines and Energy. That in itself is illegal, but a lot of infractions like this were just swept under the carpet. It will be interesting to see what else comes out when someone pulls up the carpet to take a good look.

In order to control the danger of a coal dust explosion, it is common practice to dilute the coal dust with what is called "rock dust," actually limestone that has been ground down into a fine powder. This rock dust is sprayed throughout a mine in sufficient quantity to make the coal dust inert and non-explosive. Miners from Cape Breton told me that at their mine it was nothing to use a couple of pallet loads of rock dust over a weekend. I couldn't believe that figure, because of the little bit I'd seen used at Westray. The Westray mine had only a few spots underground where rock dust was stored, with only about twenty bags at each location. Twenty bags of rock dust is only enough to cover about fifty feet of drift properly and dilute the coal dust so that it is even close to being inert. Whenever I saw rock dusting being done at Westray, it was only a light dusting, and usually a lot of spots were skipped over.

One day while I was working up in the Southwest Section, I

happened to walk up to where the boss was at. He had one of the scoops and was unloading some timbers, so I gave him a hand. As we worked, we talked, and he mentioned that the mine inspector was coming, so some rock dust had to be spread around.

I went back down to where we were raising the belt line so we could get the scoops out of the Southwest Section and worked at that for a couple of hours. We needed to talk to the boss, so I said I would go up and get him. As I got closer to where the boss was, I could see puffs of dust coming from the front of the scoop. As I walked around the scoop, I saw the boss throwing handfuls of rock dust over the transformers. I asked him what he was doing and he said, "I want to get some rock dust around before the inspector gets here." I couldn't believe what I was seeing or hearing from this guy. He actually thought this little bit of rock dust was enough to fool the mine inspectors. To my surprise it did, I guess, because not too much was ever said to the contrary.

It's simple to figure out that very little rock dusting was done at Westray. Just subtract the amount of rock dust left at the mine site from the total amount bought, and then divide that amount by the number of months the mine was in operation to get the tons per month used. Of course, after the explosion, bags of rock dust were used to build barricades and block off the Southwest Section. I daresay that probably almost as much rock dust was used after the explosion as before.

There is a way to make explosion barricades by suspending a piece of plywood at the top of a drift and then loading it up with rock dust. If an explosion occurs, the trip cords on the plywood let go, filling that part of the drift with rock dust. With that amount of rock dust in the air, the explosion is cut off and contained in only one section of the mine. It is not a save-all, but it could have saved fifteen men and a lot of damage. This method was mentioned several times to management but was just ignored. I don't know why safety suggestions were ignored so much when the heads of management were always saying mine safety was paramount.

On May 8th, I got up for work excited because it was the fourth shift, which meant I would be going home later for my days off. Everybody always seemed to be in a good mood on the fourth shift, just glad to be getting out of there. As we talked with the night crew, we found out that the methanometer on the miner in the Southwest Section was not working. As we hopped on the tractor to head underground, I told the boss what we had been told about the miner, but he just shrugged his shoulders, said "First I heard about it" and sat down ready to go. I looked at him and said, "Well, now that you have heard about it, what are you going to do about it?" He said, "Nothing I can do." I guess because it was my fourth shift, I said laughingly, "If we get killed, I'll never speak to you again." I will never forget saying that or how I feel about it now.

The tractor started up and off we went down into our little hole in the ground. Throughout the day, I asked several times about the methanometer on the miner, and our crew discussed it at lunch and whenever we got together. I was mining on D Road in the afternoon at around 4:30 pm, so I knew where the mining crew would be on the night shift.

I left work at 5:00 pm to go get my car fixed before I headed home. Whenever I left the mine site I would flip my rear view mirror up so I wouldn't have to look at it as I was leaving. I honestly hated the place that much but was really stuck there because of the work situation in the mining industry and, for that matter, every industry.

It is hard to explain to people how scary it is to leave a profession you have worked at for twelve years and look for work somewhere else. I guess all you can do is imagine what it would be like for yourself to go out and start a new career and the fears and worries of being able to make it at something totally foreign, being able to support your family and not lose our home and everything else you have worked for years to acquire.

Len later told me that the methanometer on the miner in the Southwest Section had been repaired by about noon on that last day shift. He had been told to go up and help out on the bolter with Wyman. After a while, Len noticed it getting quite warm and then he saw that the vent tube was blocked off. When he asked why it was blocked, he was told the miner needed the extra suction on its vent tube to keep from gassing out. At 1.2% methane, the miner will shut off automatically. By blocking off the vent tube at the bolter, the fan will suck bad air only from the mining face. This practice was very dangerous because the methane would build up at the bolter, which was not equipped with any means to detect the gas.

Len tore down the blocking on the vent tube, and about ten or fifteen minutes later the boss came in. He took a reading with his handheld methanometer and it read 3.75%, which is getting very close to the explosive range. The methane level must have been well into the explosive range while the vent tube was blocked. This was not the first time something like this had happened.

I got up around eight o'clock Saturday morning feeling good to be home. About ten o'clock the phone rang and I heard Mike Piche's voice. He sounded very upset. He explained there had been an explosion at the mine but he didn't have any details except that it had been massive. I hung up the phone and just wept my eyes out. I turned on the TV but didn't get much information there. I called the mine but couldn't get through, so I sat and waited for the mine to call me.

Finally, about six o'clock that evening, Rosilee, one of the office staff, called and asked if I would come in. She said I could get through the police roadblock by giving my name because she would tell them I was on my way. I left immediately and arrived at the mine about 8:45 where I was met at the front door by Arnie Smith. He asked me if I was going to go down as a draegerman and I said, "That's why I'm here."

I was not a member of Westray's mine rescue teams because I lived in Fall River and it was too far to go for practices, but I am an experienced draegerman. Arnie told me to get geared up right away because I would be going down with a team from Devco. My job was to help them find their way around the mine and be the number two man on the team. After I had finished getting into my underground clothes, Arnie brought me over and introduced me to Gary Wadden, the captain, who then introduced me to the rest of the team. As it turned out, we had already met at mine rescue competitions in 1990 and 1991. I had watched these guys in action, so I felt quite confident about working with them in this situation. We got geared up in our draegers and headed in to be briefed on our assignment.

After the briefing, we sat and waited for what seemed like an eternity until we got the OK to go to the portal to enter the mine. We got there and were told to wait in a trailer that was set up for each team to keep dry and warm while the team underground made its way to the surface. Finally we got the go from the coordinator and

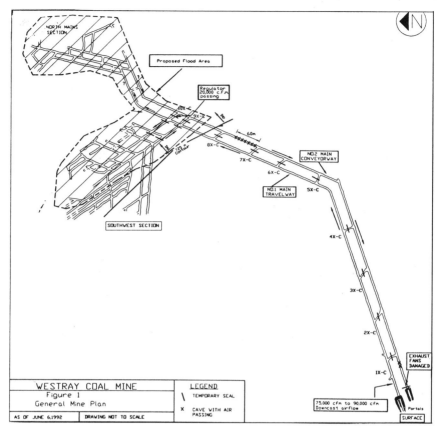

A drawing of the Westray mine underground, made after the explosion, showing the proposed flood area.

proceeded down the main decline.

I could not believe my eyes as we descended into what can only be described as hell. Sheets of metal had been blasted off the portal, as the public could see on TV, but what lay beyond the public's view was absolutely horrifying. There was a smell I can't really describe, a smell of burning mixed with the smell of pulverized rock. There was a hard black coating on the down-ramp side of all the pipes and arches. The mine had a deafening silence—no fans, no humming transformers, just nothing, nothing at all. I couldn't believe how quiet it was as we sat at No. 5 Crosscut waiting to go into No. 8.

All the walls and roof were covered with a black coating of burnt dust and debris. It was like the inside of a cannon barrel that had been loaded with anything and everything. The further down we went, the worse it looked. Cement bulkheads that had been two and three feet thick had been smashed into little pieces and thrown a hundred feet or more. Steel doors that had stood twelve and fifteen feet high and about twelve feet wide were now crumpled, twisted pieces of strange-looking metal. As we approached what I remember as No. 6 Crosscut, I saw three transformers out in the main decline. Two had been smashed together so hard that they looked like one hunk of debris. These transformers, which weigh roughly seven tons each, had been thrown about sixty to a hundred feet across the crosscut and down the decline.

From here we had to get off the tractor and proceed on foot because it was impossible to get by these mounds of steel except by crawling over, around and through them. Our job this time in was to erect a temporary seal in No. 8 Crosscut to further the fresh air into the mine. We took a two-by-four each and some tarps down to No. 8 and began to build a seal. About forty or fifty five-gallon jugs of some sort of silicon gel were lying all over the place. Some were open and leaking on the ground, which made it difficult to walk and maintain your balance while working. We eventually got the seal in place and returned to the fresh air base, made our report to the incoming team and departed for surface.

Once on surface, Gary and I went up to the boardroom where the men in charge of the operation were located and gave our account of what we had seen and done. Now it was time to wait to be called again. It was now around 5:00 am on Sunday morning. After we showered and dressed, we were to go get something to eat and to get some sleep. Sleep never did come.

About 9:00 am we were called back in. While we were getting geared up, there was a hubbub outside our room. We could hear someone saying that they needed some oxy sixties, a kind of breathing apparatus used to bring survivors out. We thought out loud, "Thank God, they have found some of the guys alive!" That was very uplifting to us all. We were all trying to find out what was going on by sticking our heads out the door and craning for information.

After we had finished getting ready and gone to our briefing, we were informed that a team had found eleven bodies in the Southwest Section and we were to go in and start taking them out. My heart was in my throat and it was all I could do to keep from crying. I now knew the job I had to do. It would not be easy, but it was definitely necessary. After some wait, we headed to the portal and then back down into what again seemed like hell, plain and simple.

As we waited to go into the Southwest Section, it was very cold and we didn't really know what to say to each other. Someone on the team told a joke, I guess to keep from going to pieces or to get our minds off what we were about to experience. It did help, but I couldn't help feeling weird about it.

As we sat waiting, a call came down from the surface for Roger. I got most of what was said and I was totally disgusted. Gerald Phillips wanted us to piggyback the bodies, meaning he wanted us to go in and drag all the bodies as far as we could, then the next team do the same thing and so on until all the bodies were out. This is one time I had to credit Roger with some dignity. He said he didn't give a shit what Gerald wanted. We would go in and remove the men with the utmost respect: one at a time and each man on a stretcher.

The time came to put on our facepieces and head in. We walked down to the No. 9 Crosscut, and I couldn't believe my eyes. Where there had been a cement overcast for ventilation, there was now nothing except a few bits of the wall lying on the floor. The roof was in fair condition, but you didn't touch anything without first checking all around you for loose or unstable areas. All the arches were knocked over like dominoes, and the conveyor belt was underneath them. Sheets of metal and six-by-six timbers had been thrown all around the drift, which made travelling very slow and extremely dangerous. There was the everpresent thought that out this close to the main decline the percentage of methane would be in the explosive range and you were walking on material that could very easily ignite another blast. All the pipes were knocked off the wall and all the electrical cables were down on the ground.

The first three or four hundred feet was the hardest travelling on our way into the Southwest Section. But once we got past that area and started up the hill towards the working areas, there was still a lot of dust and debris lying around on the roadway. One of the strangest things I saw was a set of arches twisted around like a giant wagon wheel. All of the vent tubing was gone, burnt to cinders, except for the steel wire that had spiralled along the length of the tubing. It reminded me of the razor wire the army uses to keep people out.

As we turned the corner of the top drift, I could see a few body bags lying on the side of the drift, which meant the team ahead of us must have bagged several of the men already. Further into the drift, the boom truck sat at the intersection of the top road and the beltline road. A load of rockbolts were half on and half off the flatbed of the truck. The seat of the truck was completely burnt away except for its metal frame.

As we walked around the back of the boom truck, I looked up the drift ahead of the truck and I could see some of the bodies of the second group of men lying there. I lowered my eyes and asked God to be kind to their souls and to give their families comfort and strength to get through this horrible ordeal.

When we walked into the drift where the other bodies were, I couldn't help feeling lost and completely helpless. Three men were already in body bags, so we started looking for the one marked with a number 2, which was to be the man we would take out. I recall seeing one man lying there with a blanket over his face, but I knew who he was and felt a lump form in my throat. It hurt me very much to see these men taken from us, all for the sake of greed. I took a deep breath and told myself to put it out of my mind for now and get these guys out to their families. We placed our man onto the stretcher and the team headed back out of this manmade hell.

The methane level at this point was 40% and the carbon monoxide was 7%, with only 2% oxygen—not explosive, but deadly if you were to try and breathe it. Progress was very slow on the way out, because the closer we got to the main decline, the more difficult the travelling became. You were getting tired and, because you were working in an explosive atmosphere, every move had to be calculated. Once we got out into the main decline again and placed our man up with the first man taken out, I felt relief but also despair.

We were all soaked from head to toe from our hard work in the hot temperatures of the Southwest Section. As we walked up to the fresh air base, we started to cool down very quickly in the cold air being forced into the mine. A few teams were there as backup, but the one that stands out in my mind was the team from New Brunswick. As we were giving our report to the underground coordinator, the men from this team came over and wrapped a blanket around each of us to keep us from freezing. The thing was, the blankets they gave us were the ones they had been using to keep themselves warm. After our report was finished, we headed for surface, but I couldn't

recall at that point whether it was night or day. It was about 5:00 pm
on Sunday. I didn't realize then that this nightmare was only part of
the way through.

Once on surface, we removed our gear and then had to go
upstairs and make a report to the people in charge of the rescue
operation. When the report was finished, I headed to the showers
feeling pretty depressed and angry about this whole tragic event.
Halfway across the deployment area, a guy came up to me and
asked, had we brought out Eugene Johnson yet? I began to tell him
that we were told not to say anything to anybody, when he said,
"Eugene is my brother." I don't believe there has been a time in my
life when I felt the way I did at that moment, and all I could do was
look him in the face and tell him we had taken his brother's body out
to No. 9 Crosscut. After I told him that, he put his hand on my
shoulder and thanked me for what we had done. If I had not turned
to walk away, I'm sure I would have started to cry right there in the
middle of the room.

While walking to the shower, a few of my friends came up to me
and asked how I was holding together, but I couldn't even talk at
this point, so I just kept walking. The shower was hot and it seemed
to ease the pain in my mind.

When I came out of the changing room, I noticed Gerald Phillips
talking to Lenny, so I went over to see what was going on. Lenny
told me that Phillips had asked him to go to the morgue to identify
the bodies and he was on his way. I asked Lenny if he had ever seen
a corpse before and told him I didn't think he should do this. Len and
I talked about it for a while, but I could not convince him to decline
this gruesome task, so we decided to meet back at his place later.

I planned on sticking around for a while just in case they needed
us again soon. About an hour or two passed and then we were all
called into the briefing room. After some short discussions, Reg told
us to go back to the motel and get some rest and that we would be
called as soon as we were needed. We were not doing anyone any
good hanging around the mine site, so he had to order us to leave.
But it wasn't more than a few hours before we received a call to
come in and get ready to go underground again. Once we were
geared up and briefed, we headed to the portal and proceeded
below.

Our assignment this time was to go down as a backup for one of the advance teams. We weren't down at the fresh air base at No. 9 Crosscut for very long before we could really feel the cold. Some of us built a crude windbreak out of some of the leftover tarps and we all took refuge under there, wrapping ourselves in blankets.

The two advance teams walked down as far as No. 9 Crosscut before they put on their facepieces. Glenn's team was off in a matter of moments, but Jay's team seemed to be having some kind of trouble getting their facepieces to seal properly. As it turned out, all five team members couldn't get their facepieces to seal, so we gave them ours, which left us barefaced. A call went up to surface right away to bring down five more fresh facepieces, because Jay's team could not leave until they had a backup team in full gear. Once we received our facepieces, Jay's team went on their way. About forty-five minutes later we could see lights coming back up the decline, which was very odd because each mission was usually three or four hours long. Glenn's team had travelled as far as they could before they had come to a rockfall that was impassable, and Jay's team had returned because one member had lost his nerve.

Once a member of a team is distressed, a good captain will scrub the mission and return to the fresh air base. This time down, everything that could go wrong did, and so it was decided that all the teams should return to surface and get straightened out. Once on the surface, it was easier to regroup and get yourself together, have a coffee and calm down. It was a very long time waiting for your orders to get ready, and there was a lot of confusion and arguing about what was going to go on underground the next trip in. I was getting so exhausted that I had buzzing in my ears, so I informed Arnie that I was going to have to lay down and he should get someone else to go with the team for the next run.

After looking around for a spot where I wouldn't be bothered, I finally crawled onto a wooden bench in the locker room and passed out. About an hour had passed when someone came in to change their clothes, which woke me up, so I went out to deployment to see what was going on. By this time it was about seven o'clock Monday morning and the team was just coming back from underground. Gary told me that not much had happened, just that they had been a backup for Jay and had frozen their asses off. After they cleaned up, it was off to the motel for breakfast and to try to get some sleep. With six guys in one room it was impossible to sleep, so I went over to Lenny's.

He wasn't back from the morgue yet and his wife was really worried about him. She asked me if I would go and look for him, and I told her I would check out a few places where he might be. He wasn't at any of the places I looked, so I went back to his house and his wife told me he had come in very upset and gone to bed. I knew I had to get some sleep and the only way was to get away from this nightmare, so I drive home to Fall River. This was the first time I did get some sleep, although it was interrupted by horrible visions of what I had seen and what I had to do.

Later that day my mom called, saying she wanted to go back up with me and stay at her cousin's, so we drove up and arrived at around 8:00 pm. She drove me right to the mine site and when I walked in I could see Gary and the rest of the team in the hall, talking. Gary told me that Reg wanted him to guarantee that the team would go down into the North Mains. But Gary wanted to take a look at it before he could guarantee any such thing. He said he could not tell his team to do something they didn't want to do, and he wanted a chance to assess the situation. I thought Gary was being reasonable, but the choice was guarantee or forget it, so he decided to leave.

At this point the company told the press that some of the teams were too worn out to continue or some such bullshit. At any rate, I was then assigned to another team led by Doug Rideout. Over the next few days, Doug demonstrated to me that he was what every team captain should be. I don't know if I could have kept going as long as I did if not for this man. He kept his team sharp when they had to be, and he seemed to know when to lift your spirits with a word of encouragement in a critical situation.

We geared up and were briefed, and then someone said to go, so we hopped into the truck and went over to the portal to wait for a tractor to take us down. After waiting for about twenty minutes, we were told to return to the main building, so we went back to find out what was going on. Reg asked Doug who had told us to go to the portal in the first place, and then he told us to head upstairs and wait for him in the room they had set up for coffee. When Reg came in, he had a priest with him whom he introduced to everybody. The priest told us how everyone was praying for our safety and he showed us some cards the kids of the trapped men had made for us. It was really nice to think that these poor children, whose fathers were down in that hellhole, were actually thinking of us. After talking for a while, Reg told us our team wasn't going to be going underground just yet and that we should go back to the motel and wait for a call.

We were pretty disappointed because it took quite a bit to prepare yourself mentally to go down into that place and, once you were set to go, you had a hard time calming down again. But, with no say in the matter, we left the room, had a shower and headed back to the motel to await a call. I told Doug that I was going over to Len's. After I got to Len's and we had a couple of cups of coffee, we decided to go and pay Carl Guptil a visit to see how he was holding together. As we walked in, we were introduced to two guys who were there from one of the TV stations. This reporter started asking us if we would do an interview with him on camera to tell the public what it was like at the mine site. This guy really tried to lay on a guilt trip to convince us to do this interview for him, but Len and I discussed it and told him that we were going to see someone and get some advice about all of this. Being as tired as we were, it was the smartest thing to do because it was very hard to reason or think straight.

It was about 2:30 am when I knocked on Mike Piche's door, which, as you can imagine, woke him with quite a start. We explained what was going on and how we felt, so Mike called upstairs to Andrew's room to get him out of bed for some legal advice. Andrew told us that doing an interview at this time would not be a good idea and, if we liked, he and Mike would go back to Carl's and tell this reporter to ease off. Things got pretty heated at Carl's, but eventually the reporter and cameraman left. Carl felt we had let him down, because he thought that we didn't want to tell what we knew about things that had let up to the explosion. I told him to rest assured that we would definitely be telling our story when the time was right.

We left Carl's around 4:50 am and headed back to Mike's room, dropped him and Andrew off, and then went to Len's house. It's impossible to explain how hard it was dealing with that reporter that night, but looking back I see that he was just trying to do his job.

At Len's we drank a few more coffees and then I left and went up to my cousin's house to try to get some rest. After an hour or so, around eleven o'clock, I fell asleep, and at one o'clock I woke to hear that they had got through to the North Mains and found four more bodies. Hearing that, I started to cry my eyes out, because deep in my heart I knew what the end result was going to be.

I got up and called the mine site to see when we were going to be called in. The woman in charge of calling the teams said she had no idea. I was now getting very angry at the way this whole operation was being handled. I told her to get on the ball because there were rescue teams sitting at the motel waiting to hear from her and she should at least call them and tell them what was happening. She assured me she would do that right away, so I hung up the phone and went over to the motel. I talked to Doug and the rest of the team for about an hour and then went back to my cousin's and called the mine site again. It was very difficult to sit and wait, not knowing what is going on or when you would be called back, so, needless to say, the woman at the other end of the phone got a good talking to. After she had gone to check on our status, she returned to the phone and told me to report at the mine in half an hour.

When I arrived at the mine, Doug and the team were just getting there, so we went in together and got geared up and prepared for our briefing. We were told about the men they had found, how far they had advanced, and that we were to go in and continue to explore the mine as far as we could.

Once we arrived at No. 10 Crosscut we had to change our boots to hip waders so we could get through the water that had settled just below the crosscut. There was to be no talking, no touching the walls or anything that could be left alone, and we were to walk with extreme care because there was lots of debris under the water. As I was making my way through the water, I looked up and all I could see were six-by-six timber piles up above us like a bunch of pick-up sticks. You knew in the back of your mind that if one timber came down, they would all come. On the other side of the water we had to change back into our regular boots again before we continued on. With our boots back on, we turned to go, but all we could see was a mountain of rock where it had caved in.

We went up the slope, which brought you up above what used to be the drift but was now buried. We walked along the top of this rockfall for about 150 feet before we came to one set of arches that had held the roof up. Now we had to climb down and around these arches, squeeze through an opening between the arch and the wall, and go down a ladder to the floor of the drift. From here we could only walk thirty feet and then crawl a short distance back up another rockfall before we had to climb back down to the drift floor again. From here, travel was not too bad for a long way down the No. 1 North Main. As we passed by the crosscut, I looked in, but all I could see was a wall of rock from cave-ins. This meant that all these accesses to the No. 2 North Main were completely shut off. My heart began to sink.

A small fresh air base had been set up at the No. 4 Crosscut, and immediately ahead of that was a wall of rock that had a small opening between the roof and the fall. I looked at that and knew it was the only way into the North Mains work area and we would have to go up and in there. When I got up to the top of the rockfall, I couldn't help thinking of my wife and my kids, and I said a little prayer that I would get out of this hell and see them again.

We had to travel this rockfall in the same way we had in the water, and when we climbed up and down we went one at a time because of the rocks that would fall. The roof was nothing more than layers of sharp rock above your head that could cut you or your breathing hose if it fell on you. The width of the travelway was about five feet at the bottom and it triangled up to a point at the top, like a steeple.

At the end of this fall of ground we had to pass through an opening that was just big enough to wiggle through and then half hang, half drop ourselves down to a set of arches that had been knocked over. This brought us to the beginning of a road at the level of the actual floor. As I looked right, all I could see was another solid wall of broken rock, but to my left I could see a short distance and then another rockfall. This was the way we would have to go to get into the section where the bodies had been found. This fall wasn't as large as the last one, but it was every bit as frightening. Once we made it over this fall, we were at the corner of A Road and what I called North Main No. 1 Crosscut.

The walking was pretty good from here into D Road, with only a few obstacles to go around. Just below B Road lay two flipped-over tractors, which at first glance appeared to be one. After looking at them for a moment, you could see that there were eight tires, not four. Out in front of those tractors was another tractor that was flipped over. About ten feet or so beyond that was where the first two men were laying. Doug decided we would put these two into body bags that were hunter orange so the following teams would not step on them as they came around the tractor. Doug told three of us to stay where we were, and he and two other guys went down to put them in the bags. This was when I came to the realization that we would not find anyone alive down here. As they were putting the first man into the bag, it was more than I could handle and I wept, I just wept.

After that job was done, we started down D Road, passing by the scooptram and examining it as we went. Not too far out in front of the scoop was another rockfall that went up higher than the roof, making the road impassable. We took some notes, and then as we turned to head back up D Road we heard a fall of ground. This scared the shit out of us. Just as we turned the corner at the crosscut, another fall was heard and then we felt a change in air pressure, which meant that a significant amount of rock had fallen somewhere. Every man there wondered the same thing: Were we trapped in? We discussed it and we all decided to continue with the task at hand. We turned down B Road and that is when we heard another fall of rock.

About this time we were getting really nervous, but we kept on going. Half way down B Road, just as we had gone by the continuous miner, we felt another change in air pressure. This stopped three members of the team in their tracks, so we talked about it and decided to get out. Doug, the number three man and I went down to the end of B Road to take a quick look while the others waited for us by the miner. We then started to make our way back out to where Roger should be waiting, at the No. 4 Crosscut. As we climbed up the rockfalls, I remember praying that our way out was not blocked. That was the one and only time I was glad to see Roger's face. The rest of the trip out was uneventful, except when someone on the team almost fell in the water, probably more embarrassing than hazardous.

After we reached surface and I had a shower, I went to see Rosilee and told her I would not be going back underground again. The next day was Thursday and my son's birthday, I wanted and needed to see my wife and family, and I was completely burned out. I left the mine site that day with a very heavy heart and feeling as though I had let the families down. If we could have got even one man out alive it would have made a big difference in the way I was feeling, but it just was not to be.

The next day the teams found one more man but couldn't get him out because of where he was and because the mine was becoming too unsafe to travel. The search was called off and all hope of survivors was gone. My heart goes out to the families every day that passes.

After I got home from New Glasgow on May 14th, I called the doctor's office to see if I could get something to help me sleep. The receptionist told me to get some sleeping pills and try to relax and get some rest. As I was walking down to the drugstore, everything seemed to be much more pronounced: the smells of the fresh grass starting to grow, the air so clear and refreshing, everything was so alive. It was like I had just become superaware of everything around me. I was glad to be through with going down in the mine and seeing all that destruction and death.

When I got back home, I couldn't sleep because of the phone ringing and I guess I was still wound up pretty tight. I could hardly wait to see my wife and little boy. I called my daughter who lives in Dartmouth to let her know I was home and safe. Once my family got home, I felt a lot better, and after supper I finally fell off to sleep.

The next day was the beginning of a long, hard time trying to deal with what had happened. It was a long while before I got a night of sleep without waking up with bad dreams and visions in my head. I decided to write this book to help me cope and to let the public know the truth about a company that has tried to make itself look like the victim of this whole incident.

What actually caused the massive explosion at the Westray mine on May 9th? We can't be sure without further investigations underground, but one combination of the following explanations is probably correct.

After discussing it, Len and I realized that the vent tube at the bolter had been blocked off during the day shift and was probably also blocked during the night shift. Either the methane built up at the bolter and could not be vented because of the blocked tubing, or the seal down in the old Southwest Section broke, causing an in-rush of methane. If an in-rush happened, it would not have been ventilated properly; instead of going out the return air, it would have gone up to the work face.

With no methanometer, the crew on the bolter would not have realized that methane levels were quickly rising. When the operator started to tighten or spin up a rockbolt, sparks would have been created at the contact point of the drill socket, or *dolly,* and the rockbolt. With methane levels between 5% and 15%, a smaller explosion would have resulted and then the coal dust would have ignited, creating a tremendous blast.

The way all the arches and debris were blown out towards the support mains, the blast probably started at the bolter or, if not there, at the miner. If it began at the miner, it would have resulted from a fire caused by methane and a spark from the bits on the cutting head. The fire would have travelled back over the miner and shuttle car before the coal dust ignited. I say this because when I was in the area of the miner during the recovery, neither the shuttle car, the miner nor the last thirty feet of vent tube were burnt. Either the explosion did not reach this area or it had started there. This second theory is plausible, but I personally feel very strongly that the explosion started at the bolter.

The company was very good at deception, as everyone learned after viewing what went on during the rescue attempt. I knew something was going on as soon as I saw the pretty boy they had brought in from Toronto to do all the TV reports. I had never seen this guy before. He did an excellent job of making the public feel sorry for the poor company, but it was all done that way with one goal in mind: public pity.

The first time I had been underground waiting to do my assignment on a mine rescue team showed me where the company's true concerns were. We were standing down by the three transformers that had been thrown out into the mains, and the head mechanic said, "You know, we had just put these transformers in here to keep them out of the rain. It's a bloody sin they got destroyed." When he said that, I looked at him and said, "Well, they're out of the rain, asshole." This is when my eyes began to open to what the company was really worried about.

I have never seen so much snow in May. The media was snowed, and so was everyone else. The company controlled every bit of information given out to the public and the families. The company knew for hours what was going on underground before they decided to send someone over to the fire hall to let the families know. Westray did an excellent job of turning everybody against the media. They told the rescue crews not to talk to anyone in the news because the media were just trying to get facts to twist around. Little did we know at the time what the company was up to. I'm not saying all the company staff, but a few were guilty of this manipulation. Most of the people at the mine site were genuine in their horror and concern. Most did whatever they could to help out. Some of the staff people were there for days on end, cleaning up, serving coffee or sandwiches and providing encouragement and kind words to those of us in need of it.

After the recovery operation was called off and everyone except a few people were laid off is when things started to look good for a union vote. A lot of people said that the union had no business calling a vote then, but I say they had every right to. The organizing drive had started a couple of months before the explosion, and on May 9th union organizer Mike Piche was there to help out in any way he could. He knew there would be no dues to collect and it was going to cost the United Steel Workers a lot of money to stay and stick it out. Knowing all this, Mike and the steel workers union stayed when I think other unions might have cut their losses and run.

We really had to work hard to get a vote organized quickly. It wouldn't be long before a lot of guys would be going home to Cape Breton, Newfoundland or wherever they lived. The company really helped our cause by telling whoever was left working at the mine site that if they were afraid to go down in the mine, then fuck off home. This little episode brought the rest of the previously unconvinced mechanics and electricians over to the union's side. The vote took place and the union won with 82.5% yes votes. Now we felt a lot more confident that things would be dealt with properly. It wasn't long before we saw our lawyers in action, getting set up for a major inquiry. And other unions came to our aid with very generous donations to help us get on our feet in this time of crisis.

The company started to help out a little, too, with a house to use as an office, and they set up a drop-in help centre. If a guy wanted, he could drop in at the centre and just sit and talk to someone. I talked to some of the guys who went there to find out what happened. Some said there were colouring books and free hats and other trinkets for your family, free coffee and soft drinks, but there wasn't any professional help there. When I checked into it, I was told a counsellor had been there for the first week or so but wasn't there anymore. I then found out through some people that the teams from Cape Breton and New Brunswick had received professional help from a team of crisis counsellors. This is what was known to me as "debriefing." When I went to the drop-in centre at Westray to ask about this, they told me it was in the works. Well, about a month or more later I got a call to come up for debriefing. Too little too late was the general feeling among all who attended and everybody walked out. The counsellors all had good intentions, but it was far too long a time after the incident.

I don't think many people have given much thought to the disaster's effects on the men who worked at Westray and were involved in the recovery operation. Some of these men may never be able to work underground again because of what happened. This tragic event is something you do not leave behind once it is over. The emotional wounds caused by this may never heal, and they affect more than the man who carries them. The families of these men have suffered too because of the sudden mood swings caused by what is known as post-traumatic stress disorder. I know all about this because I am one of the men who went through this, and I have seen the problems these mood swings can cause. I am finally getting my own life back under control and I am glad, really glad. It isn't easy for a family to cope with someone who can change his mood in an instant. I want to tell my family and friends how bad I feel about being that way and how happy I am that they stuck by me.

While working at Westray, I had heard talk about how Curragh Inc. had made some deals behind closed doors to acquire the mining rights to the Plymouth property. These deals were made with our so-called honest politicians. It was also rumoured that Curragh had not had the best proposal or best financial package. It escapes me why such a company would be granted the mining rights unless there was some funny business going on.

I personally think that most of those who were involved in the birth of Westray had their hands dirtied during its conception. Even as I wrote this, I noticed a news story that most of the fifty-two charges against the managers of Westray had been dropped. This proves to me that quite possibly money will yet buy freedom for those guilty of wrongdoing. I'm quite sure there will be no feelings of remorse among the guilty parties if they are freed of responsibility.

I sometimes wonder if these people are very sly at getting out of trouble or just lucky. I also wonder if they are really stupid enough to think that nothing wrong was ever done in the way the mine was run. From my experience of working there and listening to some of the ideas they came up with, I could understand them being that daft.

Every time they went to another mine in the States, they would come back saying they had the solution to the roof problems, and all sorts of changes were made to the bolting sequence and kind of ground support used. At first we were using six- and eight-foot bolts with strapping as added support. After one trip to the States, they came back and decided to try installing support beams called Birmingham trusses and to ease up on installing arches. This didn't last too long. Then we changed to another support called a Dywidag, which is a one-inch grade-60 steel bar, all threads hot rolled, which has a thirty-five-ton load capacity. In the right type of ground they are very good as roof supports, but in the ground at Westray the roof would flake and come down around the Dywidag and cave in. Although they didn't work all that well, they kept using them right up until the end. Different bolt patterns were also tried to no avail.

I can recall one week when we were mining along a flat section of No. 1 Main below No. 10 Crosscut, a short time after there had already been a major change in drift elevation. The first shift we cut, we were told to bring the floor level down a bit and the next shift we were told to bring the floor up. The following shift, after we had finished cutting, the boss came down and said that he was told to make sure the floor was brought down. I couldn't believe what was happening in this drift. The floor was beginning to look like a roller-coaster ride. When I asked what was going on, the boss said, "The surveyor and management are arguing about which elevation we should mine the floor." I couldn't understand why the company would hire a surveyor with all the know-how and experience that Ray had and then try to overrule his work.

It was as if management would awake from dreams they were having and decide to try out their dreams at work. Some of the ways they wanted things done couldn't have possibly been thought out or proven to work. Sometimes I wonder if these clowns will be allowed to run another mine and, if so, who will die as a result.

I, and the majority of the men who worked at Westray, have no respect for these guys and, as a matter of fact, I can't stand to be in the same room with them. I did have to attend a few meetings with them, and I noticed that they would not look me in the eye.

I abhor what happened to the twenty-six men and their families, the hell all of us have gone through because of this, and the possibility that maybe, just maybe, nothing will ever be done about it. It worries me to think about how many other businesses might be protected by dirty politicians, making safety rules and regulations only words written on paper. I can only hope that a government with backbone will make company officials like Westray's straighten up and run responsible and safe businesses or close their doors.

Former Premier Donald Cameron of Nova Scotia had his hands in this pie right from the start. It seems to me that he helped Curragh obtain whatever they could not get by themselves. He did this with personal gain in mind. Every move was well coordinated: major announcements were made during election campaigns and the announcement of federal support came just days before the 1988 provincial election. This deception was created to glorify the whole idea of the mine getting under way.

In fact, a report from Peter Hacquebard dated October 14, 1988, warned that faults in the area would pose a major mining problem and recommended more drilling in the area of the recoverable coal. A Canmet (Canadian Centre for Mineral and Energy Technology) report also recommended more drilling and a more in-depth evaluation of the project. It suggested that the federal Department of Industry, Science and Technology (DIST) should force Westray to disclose details about its efforts to secure commercial lending.

What effects these reports had on DIST's new minister, Harvie Andre is not clear, but it is known that the proponents of the mine were getting very worried about the delay in federal funding. Central Nova MP Elmer MacKay should have had extra clout in pushing the Westray project through in Ottawa because he was given responsibility for the Atlantic Canada Opportunities Agency (ACOA) in Brian Mulroney's post-election cabinet, but it was Cameron who flexed the muscle. The mine site was started January 1989 to make it look as if everything was going as scheduled, but in February MacKay announced a hold-up in federal funding.

Cameron and Curragh were planning a scheme to move the project ahead to the next stage, tunnel development. The province provided an interim loan of $8 million that wasn't conditional on federal funding being in place. The conditions and terms were left up to, who else, Don Cameron. Cameron saw the loan as adding to the pressure. "[If] the province goes ahead with the $8 million interim financing with Westray, then once started the project cannot be allowed to stop by the province politically!" Cameron wrote in a memo to Marvin Pelley, then an executive vice-president for Curragh. These deals absolutely boggle the mind of laypeople in our communities. To me, this all adds up to dishonesty and deception on the part of our leaders. I personally take no responsibility for these particular leaders because I did not and would not ever vote for them. I have a way of looking at people and knowing if they are good or bad, and most times I'm right. This time is no exception.

Westray officials were warned time and time again about the dangers of mining this very gassy coal seam. Instead of being extraordinarily careful, they were lackadaisical about safety procedures. All anyone needed to do was read about the death toll this coal seam had already taken to know of its dangers. I don't recall an organized safety meeting ever being held the whole time I worked at Westray. Safety meetings are a very important part of the work environment at a mine because they reinforce the necessity of working safely when down underground. It is important to think safety every day you work if you don't want to get yourself or someone else hurt.

Companies such as Curragh must realize that money saved by cutting corners can be very costly in the end. If you skimp and fail to do something correctly the first time, you usually have to go back and fix it up properly at a greater cost. Management at Westray was notorious for doing things in a half-assed way and then not fixing them until something happened.

Many times after an intersection was cut, there were no arches set at the corners. It is very important to stabilize the roof and walls of an intersection because the ground is being opened up wider than the drifts. Without arches, the roof and walls of an intersection will cave and sluff off even if they are bolted. If arches are installed as soon as possible, cave-ins are reduced significantly. Once the intersection caves, it can't be trusted even after it has been arched. Almost every intersection that wasn't arched at the time it was cut was found caved-in after the explosion.

These cave-ins blocked us from getting into where the rest of the men had been working. In fact, almost every area that was caved-in by the explosion had not been arched right away after being cut, which proves my theory of the greater cost of not doing things properly the first time. I can only hope that the people in charge of the mine, should it ever open again, understand the importance of doing things right the first time.

Keeping records of incidents is another very important practice for controlling the possibility of a major accident. In the mine safety and loss control course I took, it was stated that for every 600 incidents with no apparent injury or property damage, there are thirty property damage accidents, ten minor injury accidents and one major injury accident. If records had been kept properly, management would have been able to see a very definite downfall in safety at this mine site. Very few incidents or minor injuries were ever reported in writing and very seldom, if ever, were they considered important.

Once in October, I twisted my back while climbing down from the canopy of the bolter and I was off for two weeks. When I got back to work, I was told to go up to the industrial relations office so an accident report for the workers' compensation board could be filled out. It was absolutely ridiculous that they waited until I returned to work to fill out a form that should have been completed and sent out within five days of the accident. Because of this lapse in filing reports, I had no money coming in for three weeks. As anybody who works for a living knows, this can be a real pain. Maybe someday they will realize that a happy worker is a productive worker. You work harder if you feel you are part of a team striving for the same goals as the company you work for.

The morale at Westray was low at the best of times, mainly because of the way we were treated by management. It is pretty pathetic when the mine superintendent, the second highest man at the mine site, grabs himself by the crotch and starts yelling at you to suck his cock and calling you a brain-dead fucker. This was his usual way of getting his point across to you.

I can't understand how they expected the workers to feel any pride or satisfaction working for someone like this guy. Men were afraid to make any kind of mistake at all when this man was around. At times when you try too hard not to do something wrong, you usually do. When a mistake was made, this guy was right on you. He would go up one side of you and down the other, yelling and spitting chewing tobacco all over the place. If it hadn't been so aggravating, it would have been kind of funny watching this guy going off like a crazed maniac. Imagine what your workplace would look like with gobs of chewing tobacco all over the place, and how you would feel after some guy had just chewed you out royally. I don't think you would feel much like working to make him look good in the eyes of upper management. I dare say, most people would just say "Up yours, pal" and slack off so he would look bad.

This happened quite a bit at Westray when guys got chewed out. Most of the men worked hard for the best part of the twelve-hour shift. But if they got a blast of abuse, they would slow down and sometimes stop working and complain to each other about the way they had been treated. This abusive treatment was tolerated because of the lack of other jobs to go to. It reminded me of "The Molly McQuires," a movie about miners in Ireland years ago. When I saw that show, I thought nobody would ever put up with such things nowadays, but I guess I was wrong.

I know I sound like I'm crying the blues here, but I'm just trying to explain how it was at Westray. Perhaps it will change if and when they reopen, and if it does change there will be a big change in worker attitude. Before it was the attitude of "Why should I give that extra effort if nobody even gives a shit?" That was the way it seemed to be. From conversations I have had with the people in charge of the mine site later, it seems that things would probably be run differently.

Since the explosion, the company and government have done a lot of things that are hard to understand. When it was announced that the mine would be flooded I thought, *Why would they do this?* They said the flooding was necessary to stabilize the ground conditions. In the fall of 1992 we had a meeting with Westray's ground control consultants so they could explain what they thought about the ground conditions and the best support system to use. At this meeting it was explained to us that the rock on the roof would expand and break up when water was introduced. This was the reason it was caving in when we cut new ground. The water sprays on the continuous miner were causing this shale-type rock to expand and cave. If this was the case, it would be ludicrous to flood the mine to increase ground support. This action will cause more damage than good.

The water will also wash all the coal dust down the declines to settle. If the mine is drained, the coal dust will become like quick sand or thick black glue. It will be very difficult, if not impossible, to travel through. If any of the unrecovered bodies are in this black ooze, they will more than likely not be found. I don't believe the plan to flood the mine was a plan that was carefully thought out before it was implemented. Maybe it was thought to be the best thing to do under the circumstances, but more thinking was definitely needed.

When I heard the RCMP were going into the Southwest Section, I was excited that there would be some concrete evidence to prove what had happened. To the date of writing this book, the only things announced were that the continuous miner was left in the on position and the RCMP had taken out the methanometer to analyze it. This was announced on January 17, 1993. It also seemed to take a long time for the RCMP to decide to lay charges against Westray and its managers. The provincial inquiry went absolutely no place, because the seven management personnel put forth an injunction to stop it because it was too broad in scope. They were afraid it would be unconstitutional to proceed as it was set up. To this date, there has not been any word of when or if the inquiry will start. I would like to see it begin soon, so we could find out what happened and what caused it to happen. I think a lot of questions will be answered and then it will be obvious who is at fault for this tragedy.

The officials at Westray proposed to further flood the mine up to the No. 8 Crosscut. This will block any further attempts by investigative teams to enter the Southwest Section. The reason given for flooding this time was to stabilize the mine up to the No. 8 Crosscut so a reopening of the mine could be considered. The water was to trap any gasses in the Southwest Section so work could safely begin on the new mains. But what will happen when the mine is drained so the recovery of the bodies can be done? The methane could find its way to the breakthrough point and then you will have a very dangerous situation once again. Of course, if no real attempt is ever made to recover the bodies, there won't be anything to worry about.

The families group tried to stop the flooding and were successful for a while, but a ruling handed down from the court gave Westray the go-ahead to flood. Once again the victory went to the Goliath of this story. Only time will tell what will come of all of this. When Curragh submitted a request to reopen the mine, it was met with protest from the families group and several of the miners. Most people do not realize that this company is not the only game in town. If the mining permits are not issued to Curragh, I am sure a few other mining companies will be chomping at the bit to take over this property. Because of the protest by the families and the growing dislike for Curragh's tactics, one very nasty comment was released to the press: Do they want the bodies or not? I don't think this comment was ever meant to be heard, but when the pressure is on, it is easy to speak without thinking. I hope this sort of comment will not be repeated again. It only shows a lack of concern for the families and their plight.

It came as no surprise to me when I heard that Curragh wanted to work the strip mine in Stellarton, but what was hard to believe was that there were no plans to use the laid-off miners to do the work. If the union had not stepped in and protested, the company was going to use a contractor to do the mining and only a few of our men to work at the wash plant and do odd jobs around the mine site. After much discussion, and threats not to support the strip mine, the company agreed to use as many of its own workforce as possible at this open pit. This was, as it turned out, a waste of time because an environmental study was ordered by the provincial government before further mining would be permitted. I personally agree with doing the study, because it should be clear what the mine will do to the area after it is finished. If everything looks good in the reports, then it should be permitted to go ahead. However, all the work should be done with an accent on safety and all guidelines should be followed to the letter. Curragh put up quite a fight to override the ruling for the study but later conceded and decided to wait for the completion of the study.

According to a news release of January 19, 1993, the inquiry was to be on again, but it will be a long time before the public sees any proceedings. They said that any criminal charges laid by the

RCMP must be dealt with first before the inquiry can get under way. This was not really good news because the criminal charges could take several years to be dealt with. It was like a conspiracy was being put together to cover up the whole ugly mess. Perhaps the people responsible for this think that in time we will forget about it and then it can be dealt with quietly. I, for one, and for sure the families group, will not forget. When the time comes to get up on the stand and tell what I know, I will definitely be there.

With the death toll in the Pictou coal fields now at least 270, we must strive to make changes and add regulations to a very outdated Coal Mines Regulation Act. Here are some recommendations:

1. All equipment underground in a coal mine should be equipped with a methane detection device that will shut the machine off at a predetermined setting.
2. Much higher fines should be imposed for mining infractions, for example, a $10,000 minimum.
3. If conditions are not up to specifications, the mine should be shut down immediately until the situation is rectified.

As I wrote this book, I felt many different emotions—sadness, bewilderment, anger, hurt—and a flood of different feelings. As I get close to the end of my story, I have a feeling of satisfaction, knowing that I have gotten the truth out so people will understand about life at the Westray Mine, "Satan's playground."

If and when the mine reopens, I say this to the men who go down to work: Be safe, be tough and don't compromise your right to a healthy and safe workplace. Don't be bullied or fooled into doing anything you feel is unsafe.

IN THEIR MEMORY

TWENTY-SIX ANGELS

There are twenty-six angels in Heaven today,
buried in the ground of a mine called Westray.
Fifteen were recovered and one left behind,
a guardian angel for those left in the mine.
The country is shaken, the families in shock,
people are crying, unable to talk.
Our hope has been taken by men who should know,
the men who have worked in the ground deep below.
The decision is heartbreaking for the men in the mine.
"Shall we go on or leave them behind?"
We, as the people, our emotions run high,
we can't understand why our men had to die.
The only thing left for us to do
is pray that God will get us through,
and someday find comfort that those who are loved
are watching from Heaven as angels above.

We Will Remember Them

John Thomas Bates, 56
Larry Arthur Bell, 25
Bennie Joseph Benoit, 42
Wayne Michael Conway, 38
Ferris Todd Dewan, 35
Adonis J. Dollimont, 36
Robert Steven Doyle, 22
Remi Joseph Drolet, 38
Roy Edward Feltmate, 33
Charles Robert Fraser, 29
Myles Daniel Gillis, 32
John Philip Halloran, 33
Randolph Brian House, 27

Trevor Martin Jahn, 36
Laurence Elwyn James, 34
Eugene W. Johnson, 33
Stephen Paul Lilley, 40
Michael Frederick MacKay, 38
Angus Joseph MacNeil, 39
Glenn David Martin, 35
Harry Alliston McCallum, 41
Earl Eric McIsaac, 38
George James Munroe, 38
Danny James Poplar, 39
Romeo Andrew Short, 35
Peter Francis Vickers, 38

—*written by an anonymous author*

Bennie Joseph Benoit

A wonderful, loving husband and father who is greatly missed. Remembering you is easy; missing you is a heartache. Wonderful memories of love and happiness will be held within our hearts forever. The time we had together was so very special, and your life accomplishments made us all so proud.

You are in our hearts and deeply loved forever, never to be forgotten.

—wife Shirley; daughters Kelly Ann, Nadine and Lisa; grandson Garrett; son-in-law Dana

Wayne Michael Conway

More than a year has passed since the May 9th explosion that completely tore apart our lives. Slowly we are trying to pick up the pieces and put our lives back together. It is one of the most difficult tasks we have ever had to do. Not a day goes by that we don't think of Wayne and wish he was still here with us. For twenty years he was the one constant in my life—my husband, lover and, most of all, my best friend. There were many things he asked of me over the course of our marriage, and I tried my best to accomplish everything he asked of me. But there is one thing I have not been able to do for him. About fifteen years ago, in the course of conversation, he made me promise that if he ever got hurt underground, I'd make sure they got him out of the mine. Little did I think that would ever happen. Even at the time of the accident, I never once thought we wouldn't find him. And now more than a year later, we are still no closer to recovery.

No one will ever know how hard it is not to be able to carry out that most important promise that I made.

I'm sorry, Wayne, and I love you.

—written by Wayne's loving wife, Shirley

Robert Steven Doyle

Robert "Robbie" Doyle, at the age of 22, was the youngest man to die in the Westray explosion. He was the youngest of four sons of Marshall and Marie Doyle of Plymouth. Two of Robbie's brothers also worked at the mine site: Jim, a diesel mechanic, and Allan, who was working in the wash plant the shift of the explosion.

Robbie loved life, was a friend to everyone, loved people, and was always ready to lend a helping hand. He was the one who always made us laugh. He took great pride in being a member of the Plymouth Volunteer Fire Department. Robbie loved darts and enjoyed snowmobiling.

Animals were a love of Robbie's and he spent most of his spare time on the family beef farm next to the mine property, where he spent many happy hours with his father and brothers. Robbie will never be forgotten for everything that he was, and he is sorely missed by his family, friends and community.

Robbie's great grandfather Pat Doyle of Westville was a draegerman during the Moose River mine rescue in 1936.

—written by Robbie's mother, Marie Doyle

Roy Edward Feltmate

Roy was born on August 30th, 1958, at Goshen, Nova Scotia. His parents are Wesley and Ivy Feltmate. He was the second son in a family of eight, and had three brothers and four sisters. He was educated at Goshen Elementary and at St. Mary's Rural High School, in Sherbrooke. He worked in the pulpwood industry, mainly as a "porter" operator.

Roy married (Mary) Bernadette Crispo of Monastery, Nova Scotia, and they have two daughters: Amy Lynne, who had her eleventh birthday on May 9th, 1992, and Holly Anne, who was born on August 22nd, 1984.

Roy started as a miner at Forest Hill gold mine and then on to Gays River in 1989, moving his family to Stewiacke in June 1990. He lost his father in November 1990. Roy went to work at Westray in November 1991.

Roy was a big, quiet, fun-loving man. Blessed with great strength, he was a good person to have around in a crisis. He served as Fire Chief at the Goshen Volunteer Fire Department for a term. An excellent driver, he thoroughly enjoyed stock-car racing. He played a good game of ball and was an avid hunter.

We will miss him.

—written by his mother, Ivy Feltmate

Myles "Sparkie" Daniel Gillis

Myles worked as an electrician at Westray at the time of the explosion on May 9th which took his life. He was 32 years old and married, with three children he adored.

Myles was a very kind and caring individual. He lived his life trying to do good for others. He was an active member of our local volunteer fire department and a participant in mine rescue training and competitions. He enjoyed hunting, fishing, scuba diving, and camping with his family.

Myles greatest joys came from spending time with his family. As a father, he was second to none. Our children are now forced to continue on in life without their Dad, but Myles made sure that their memories of him would be wonderful, as he always made them feel loved and well cared for. When they speak of their father today, they can't help but smile.

As for myself, I couldn't imagine having a more wonderful husband. In our short time together, Myles gave me more love, strength and courage than most people acquire in a lifetime. It seems sad that such a loving person should be taken away. Although he left us in body, his spirit remains with us. I am certain he is watching over us and acting as our guardian angel from heaven above.

To Myles: We love you and miss you. "Till we meet again."

—Isabel, his loving wife; Christopher, Ashley and Daniel, his children, his Mom and Dad, Joe and Eileen Gillis; and his brothers and sisters, Brian, Heather, Rosie and Tina

Trevor Martin Jahn

Trevor, more commonly known as "T.J.," was born on June 26, 1955, in the small mining community of Coleman, Alberta. He completed his schooling and then began his mining career at Vicary, an underground mine north of his hometown. Ferris Dewan, his lifelong friend, also grew up in Coleman and they were inseparable. Together they worked in underground mines from the West to East Coast, including Nanaimo and Grand Cache.

Known for his fun-loving spirit, T.J. knew how to work and play hard. In his spare time, he enjoyed hockey, motorcycling and working on his 1969 Camaro. In addition, he had two cats by the name of Casey and Finnegan. These cats were his companions for thirteen years and travelled with him from coast to coast.

In 1991, T.J. left the Grand Cache mine for the promise of a better life at Plymouth, Nova Scotia. With his girlfriend Bonnie, her son Jesse, and his two cats, he set out for the East Coast.

Trevor often spoke of the difficult conditions at the Nova Scotia mine but remained there in order to get sufficient funds to move elsewhere. In May 1992, disaster struck the mine, and Trevor and Ferris were among the twenty-six miners killed.

He is survived by his daughter Kristy, wife Bonnie, son Jesse, mother Reta, his brothers Larry (Francis), Marvin, Tracy and his sisters Faye (Rick), Dana (Eugene), Cheryle (Randy), Norma (Roy) and Beverly (Frank). He is also missed by many nieces, nephews and friends.

—written by Reta M. Jahn, his mother

Michael Frederick MacKay

Mike was full of fun. He had the gift of gab and the gift to make people laugh. He was always singing, whistling or talking. He was full of wit, charming, and someone you just couldn't help but like. Mike loved motorcycles. He had different models through the years, but he never did get his Harley-Davidson. He would have bought it in the spring of 1993 and was so excited to know he was getting close to his goal. The last year of his life he slept, read and breathed Harley-Davidson. That was the first thing I thought of when Mike died. I felt so bad that he was so close to getting his dream bike. You see, Mike always said, "If I didn't have bad luck, I'd have no luck at all." I guess he was proven right. I'll never forget those words. I wish I had a headstone to carve them on.

I was 14 and Mike was 19 when we first met. We were married when I was 21 and Mike was 25. We were together for twenty years. We have two daughters, Sara, 10, and Janelle, 9. Being only 14 when I met Mike, I cannot imagine my life without him. Now that I have no choice, I draw on the strength he always showed me. Mike was a beautiful, warm person who taught me a lot and he is very dearly missed and loved, not only by me and our girls, but by everyone who knew him.

—written by Mike's loving wife, Beverly

Glenn David Martin

Glenn was born in New Glasgow on May 11, 1956. He was our son and our friend. He had a host of friends who loved him and will miss him. He was a good man who worked hard and got a lot out of life. Glenn spent his leisure time in the woods, which he loved.

—written by Albert and Jean Martin, his parents

Harry Alliston McCallum

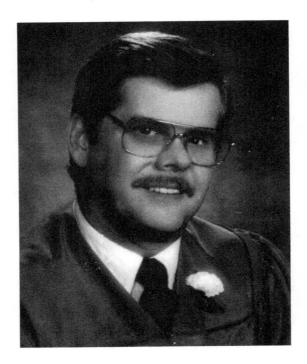

IN DADDY'S NAME

I see him still just standing there,
his figure strong and bold,
the man that I will always love
and cherish till I'm old.

We had so many special dreams
the two of us would share.
Now they're filled with time and space,
with dreams I cannot bear.

To hold him close and feel his touch,
the love of his embrace,
the words he didn't have to say
were written on his face.

Two loving children we would bear
to carry on his name.
Their futures we had hoped to mould
would never be the same.

The questions asked are ones of doubt,
Why did our daddy leave?
He wasn't here for very long,
our hearts are left to grieve.

Our memories will sustain us
though the nights are long and still.
I sit and gaze and wonder:
Is this really of God's will?

No matter what the future holds,
there's one thing that's for sure,
the one true love that people share
is love that will endure.

In memory of the only man
I will ever love, Harry.

—written by Shelle McCallum, his loving wife

Earl Eric McIsaac

In loving memory of a dear father and brother who passed away on May 9, 1992, in the Westray mine explosion at the age of 38.

Eric was born the son of a coal miner to James and Hazel McIsaac on June 19, 1953, in the small community of River Hebert, Cumberland County, Nova Scotia.

He had worked different types of jobs over the years, including coal mining at the Cobequid coal company in River Hebert for a period of time. Good jobs were scarce, and he consequently went back to coal mining when Westray opened up.

Eric was a man of simple pleasures, such as music, hunting and fishing. He enjoyed life and brought joy to a lot of people with his carefree, happy way. Eric will remain in the hearts and memories of his family and all those who knew and loved him.

—written by Mrs. Sharon LeBlanc, his sister

Romeo Andrew Short

Loving you was so easy
that letting you go
was the hardest thing
I ever had to do.
Beautiful memories will
always keep you alive
in my heart.

Romeo was a very wonderful man. He was a very devoted husband and an excellent and loving father to our two children. Romeo loved life and everything that life had to offer. He appreciated all the little things that most of us take for granted. This was sometimes due to the fact that his job took him away for a couple of weeks at a time. Sometimes he didn't see daylight for the entire two weeks that he was away because he was working the day shift for fourteen consecutive shifts.

Romeo was a quiet person who enjoyed the more simple things in life. He loved to read, he was an avid sports fan, he enjoyed politics immensely and he loved to putter around in his shed. Romeo was a man who gave much but asked for so little in return. He will be forever missed by all who knew him.

—written by Cavell Short, his loving wife

The Boys of Westray

The Albion, the Foord, we'd seen them all,
but to Westray many would answer the call.
From England, Scotland, Cape Breton and the West,
Westray asked for only the best.
The coal history of Pictou spoke of the gassy seam Foord,
but once the dust is in the veins, the lives of the men is left to the
  Lord.
The safety of mining here was a query,
the history of methane made us all leary.
In August '91, I received a call to help prepare the rescue crew,
in hopes the fears would never be true.

The evil methane lurked below,
but the men we met were a pleasure to know.
The public was at the open house, as the fears ever present,
but time passed by and the future seemed pleasant.
Training continued and new faces came,
no one aware of the lives it would soon claim.
Plans for a course on the books there had been,
but the fate of May 9th was yet to be seen.

The people of Plymouth were rudely awoken.
At 5:30 am, a blast and lives taken.
The fears, the warnings, had been foretold.
The evil gas had taken miners of old.
The boys had gone down on the night shift crew,
but what awaited them there no one knew.
Twenty-six lives were taken that day.
The sorrow, the fear, left us only to pray.
The draegermen came from all around,
in hope to bring them alive from the ground.
On Sunday the word was to be bad,
which made our community both mad and so sad.
Eleven bodies found, hopes faded for the fifteen more;
as the days dragged on, the news would be as before.
Thus came more bodies and rivers of tears,
the explosion was as fatal as our fears.
Then came word the search was no more,

the dangers below were worse than before.
Rockfalls and methane were always there,
to risk more lives would not be fair.
A cross was erected above the grave,
where lie the men they could not save.
Five of the boys we trained were gone,
but with the rest of the world we have to go on.
The memories of the guys who were part of the team,
the crew of mine rescue had made Arnie's face gleam.
They bravely worked in a treacherous place,
each day they came up with coal on the face.

The tears and the sorrow and feelings for the lost
are just a margin of the unmentionable cost.
To not say goodbye to the families and friends,
the question of "Why?" will be one that never ends.
This disaster had an effect on us all,
but once again we know a miner will answer the call.
As days go by we must put the pieces together.
Each of us has had a storm to weather.
To return to Westray was a difficult thing,
thoughts of the men and the memories it does bring.
To walk the halls and hear the tales,
their bravery and courage forever prevails.
The memories of the explosion will always live on,
even now with the brothers gone.
The feelings of their presence is quite strong,
did they know that something was wrong?
For their memories we must hope and pray
that work and searching will continue someday.
But until that day when we meet again,
dear brothers and friends, you'll be in our memories until then.

We'll never forget you, forever.
Goodbye.

—written in July 1992 and dedicated to the memories of the twenty-six men who lost their lives in an explosion on May 9th, 1992, at Plymouth, Nova Scotia, by Diane Langille, Mine Rescue–First Aid Instructor and friend of the miners

A Glossary of Mining Terms

arches: Made of heavy I-beam steel, they come in three sections that are bolted together to form a large arch. Once one arch is up, the next is attached to the first with a spreader bar about three feet long. After several arches are up, they are timbered on top and at the sides to stabilize them. When installed properly in good ground conditions, they provide a strong support for the roof.

bolter: The bolter is a complex piece of equipment used to drill bolt holes and install roof support systems. It is driven on tracks like a continuous miner and is about the same length but not as heavy. At the front end is a large T-shaped hydraulic-jack post called the "TRS" or "temporary roof support." It extends up to support the roof while rockbolts and screen are installed. Behind the TRS are two independent drilling platforms. Although separate, if one is shut off, the other will also shut off. The back end of the machine is used to store bolts and other gear used for roof support. The bolter is good for bolting but very awkward to move from one workplace to another.

chocking: To a hardrock miner, this term is "timbering," because that is exactly what you are doing. Chocking is timbering around and above the arches with what a coal miner calls a "chock block" and a hardrock miner calls a "six-by-six" timber.

continuous miner: An electric-hydraulic mining machine. It is approximately thirty feet long and weighs roughly sixty-five tons. It runs on tracks much like those on a bulldozer. The cutter head is ten feet, eight inches long and about six feet around. The many cutting teeth on the head chew up the coal and drop it onto a large spade at the front of the miner. In the spade is a set of gathering arms that pull the coal into the conveyor, which carries the coal out the back end of the miner. The back end is called the tail and that is what it looks like; it moves from side to side and up and down to put the coal where you want it. The head moves up and down on the face and chews off the coal. The operator sits in a protected cage and controls the miner with relative safety from large pieces that may come at him. One drawback of the operator's compartment was that smaller rocks could get through the screen and strike you in the face. The miner is equipped with a methanometer that will shut the machine off when a concentration of 1.25% methane is reached. The continuous miner is generally a very nice piece of equipment to work with.

crosscut: A tunnel that comes off a main drift.

deeps: The two main tunnels of the mine. No. 1 Main deep was the entrance to the mine for fresh air, men, equipment and supplies. No. 2 Main deep was the exit for bad air and the coal via conveyor belt. Hardrock terms for deeps are "ramps" or "declines." A coal miner sometimes also calls them the "mains."

dolly: When a rockbolt is tightened, a dolly is used. It is a 1 1/4-inch-square socket that is attached to a drill steel. The steel fits into a drill chuck, and the drill is used as a tightening tool.

dosco: The dosco is a track-driven mining machine, used mostly in areas with rock in the face. It has a pineapple-shaped cutting head that moves from side to side and up and down. The dosco advances more slowly than a continuous miner, but it is powered in the same way and will cut a rounded drift.

drifts: Tunnels that make up the travelways of a mine. They are called by different names according to their locations and purposes. Hardrock miners call them one name and coal miners call them another. At Westray there were quite a few hardrock miners, so conversations contained a mix of mining jargon.

face: The far end of the drift where the mining occurs.

flatbed truck (or **"boom truck"**): A diesel-powered unit used to bring gear and supplies into the mine. It has rubber tires and is about twenty-five feet in length. This machine articulates in the middle to aid steering. It has a large flatbed on the back and a hydraulic boom to load and unload heavy objects. The flatbed truck was not very practical at Westray because of the steep grades and the amount of coal dust that built up on the roadways. A lot of time was spent trying to get the truck back up the declines once it had dropped off its load. The flatbed trucks at Westray were not equipped with underground exhausts and had open starters. Both of these things were contrary to the coal mines Act.

methanometer: An electrically operated instrument used to detect and accurately measure the amount of methane gas present in the atmosphere of the mine in a range of 0–5%. Concentrations of methane over 5% were indicated on a handheld methanometer, but without the accuracy of high-range monitors. Only the scoops and the continuous miner were equipped with methanometers at Westray.

mucking out: Removing broken rock or coal with a scooptram. Scoops were not really used for their proper function at Westray, but they are invaluable in a hardrock mine.

rib: The sides of the drift if you are talking to a coal miner. A hardrock miner calls it the "wall."

rockbolts: Rockbolts come in different sizes and designs. The most common type at Westray were the re-bar type that are anchored into the rock with a resin compound. They are designed to lock the different layers of rock together and a plate on the end holds the screen in place against the roof.

roof: Just as the name suggests, the roof of a drift. The word "roof" is used by coal miners, but hardrock miners call it the "back."

scooptram: A diesel-powered machine properly used to load, haul and dump material like a front-end loader. A "scoop" is twenty-five feet long and weights thirty tons. It articulates in the middle to steer and is operated from the left side of the back end. The scoops at Westray were used mainly to move materials and gear around or as a working platform while putting up arches or chocking above them. There was a methanometer on each scoop, located on the wheel cover over the left front tire.

screening: Four-inch-square mesh used to support the roof. The screen at Westray was eight feet wide and twenty-two feet long.

shuttle car: This piece of equipment is electric-hydraulic, but it is mounted on rubber tires and can move quite a bit faster than a continuous miner. It is used to haul coal from the working face to the Stamler. A shuttle car is about thirty-five feet long and weighs roughly twenty-five tons. It has a conveyor built into its belly to remove the coal. The operator sits at the rear end, to the right or left, depending on the car. There are two seats facing in opposite directions so the operator can always face in the direction he is driving. There are dual controls for the brake and throttle, and an emergency brake bar is placed right where your arm sits. The steering control is an extremely touchy joystick. All the electric gear has a trailing cable that (except for the continuous miner) feeds out or retracts as you drive along. The hardest things to get used to with this machine were its jerky starts and its four-wheel steering, which made it difficult to turn.

Stamler: The receiving bin for the coal being emptied out of the shuttle car. A conveyor along the bottom of the Stamler moves the coal past a small breaker wheel that will smash any larger pieces of coal into smaller sizes. The coal goes out the back end of the Stamler onto a conveyor belt that takes it up to surface. As the mine advances, the Stamler conveyor belt is extended.

vent tube: A long tube of reinforced plastic used to remove bad air and dust from the working face. The tubing at Westray had a steel wire spiralled around it to keep it round and to prevent it from collapsing under the suction of the fan.